Lecture Notes in Mathematics

Edited by A. Dold and B. Eckmann

528

James E. Humphreys

Ordinary and Modular Representations of Chevalley Groups

Springer-Verlag
Berlin · Heidelberg · New York 1976

Author
James E. Humphreys
Department of Mathematics and Statistics
University of Massachusetts
Amherst, Massachusetts, 01002
USA

Library of Congress Cataloging in Publication Data

Humphreys, James E
   Ordinary and modular representations of Chevalley
groups.

   (Lecture notes in mathematics ; vol. 528)
   Bibliography: p.
   Includes index.
   1. Chevalley groups. 2. Representation of **groups**.
3. Modules (Algebra) I. Title. II. Series: Lecture
notes in mathematics (Berlin) ; vol. 528.
QA3.I28 vol. 528 [QA171] 510'.8s [512'.22] 76-19111
ISBN 3-540-07796-0

AMS Subject Classifications (1970): 20 CXX, 20 GXX

ISBN 3-540-07796-0 Springer-Verlag Berlin · Heidelberg · New York
ISBN 0-387-07796-0 Springer-Verlag New York · Heidelberg · Berlin

CONTENTS

INTRODUCTION

This paper deals with some aspects of the modular representation theory of Chevalley groups and their twisted analogues, notably the interplay between irreducible and projective modules, and then indicates using Brauer's theory some connections with the ordinary representation theory (over $\mathbb{C}$). The results in this latter direction are somewhat fragmentary, but still highly suggestive. The ordinary representations have some of the flavor of infinite dimensional representations of semisimple Lie groups, while the modular representations for the prime p defining the group reflect the influence of the Cartan-Weyl theory of highest weights for finite dimensional Lie group representations. So in effect we are trying to imitate the procedure of Harish-Chandra in looking at the "infinitesimal characters" associated with certain families of unitary representations.

Let us outline briefly our program and establish some notation. Denote by G a simply connected simple algebraic group (Chevalley group of universal type) over an algebraically closed field K of prime characteristic p. Let $\Gamma = G(\mathbb{F}_p)$, the finite Chevalley group of type G over the prime field, and let KΓ be its group algebra over K. (The letter Γ is used here to make a sharp typographical distinction between the finite group and the algebraic group; in the literature G may instead be denoted by a script or boldface letter, or else Γ may be denoted by G_σ to indicate that it consists of the fixed points under an endomorphism σ of the algebraic group, here the Frobenius map.)

The irreducible KΓ-modules M_λ are indexed by a set Λ of "restricted" highest weights, and to each corresponds a principal indecomposable module (PIM) R_λ having M_λ as its unique top and bottom composition factor. On the other hand, over a splitting field of characteristic 0, Γ has irreducible modules $Z_1, \ldots, Z_s$ which can be "reduced modulo p" to yield $K\Gamma$-modules $\bar{Z}_i$ whose composition factors are well determined. We write $\bar{Z}_i \underset{K\Gamma}{\leftrightarrow} \Sigma M_\lambda$ to indicate that the two $K\Gamma$-modules have the same composition factors, counting multiplicities (and similarly for other groups or algebras). It is known that $R_\lambda \underset{K\Gamma}{\leftrightarrow} \Sigma \bar{Z}_i$, where $\bar{Z}_i$ occurs as often as M_λ occurs above. So in a sense the ordinary representations Z_i are caught between the M_λ's and the R_λ's.

Next let $\underline{g}$ be the Lie algebra of G, and let $\underline{u}$ be its restricted universal enveloping algebra over K (of dimension $p^{\dim G}$). The representation theory of $\underline{u}$ parallels closely that of $K\Gamma$. Indeed, Curtis and Steinberg showed that the M_λ ($\lambda \in \Lambda$) are also G-modules, whose derived $\underline{u}$-modules (still denoted M_λ) are precisely the irreducible $\underline{u}$-modules. Work of the author and D.-N. Verma (for which proofs appear below) shows that the corresponding PIM's Q_λ of $\underline{u}$ are also G-modules, whose restrictions to $K\Gamma$ are projective modules involving the respective R_λ as summands (with Q_λ often but not always equal to R_λ). The analogue here of $\bar{Z}_i$ is a p^m-dimensional $\underline{u}$-module Z_λ ($\lambda \in \Lambda$), m = number of positive roots of G, characterized as the largest $\underline{u}$-module generated by a highest weight vector of weight λ. Although neither $\bar{Z}_i$ nor Z_λ can be viewed naturally as a module for G, these modules seem to have strikingly similar features (in the generic case). In particular, we can predict to some extent what the degrees and the composition factors of the $\bar{Z}_i$ ought to be, by "deforming" the Z_λ systematically. In this way we are led to associate with Γ a complex of dimension ℓ (= rank G) which generalizes the Brauer tree of SL(2,p). Although the "regular" series of representations of Γ are coming to be well understood (especially through the recent work of Deligne and Lusztig), the modular viewpoint may contribute something to our understanding of the degenerate cases: every $\bar{Z}_i$ (no matter how degenerate) does occur in some R_λ.

A common thread running through the topics just outlined is what might be called a "decomposition" problem: Determine the composition factors M_μ which occur in the G-module $\overline{V}_\lambda$ or the $\underline{u}$-module Z_λ or the $K\Gamma$-module $\overline{Z}_i$. We have already suggested a connection between the second and third of these. Recent work of J. C. Jantzen leads us to believe that there is also a remarkable connection between the first and the second (or third), involving nonrestricted weights λ.

The modular representation theory of $\Gamma_n = G(\mathbb{F}_{p^n})$ largely reduces to that of Γ, thanks to Steinberg's twisted tensor product construction. On the other hand, results about the ordinary representations of Γ_n generally come from results about Γ simply by the substitution of $q = p^n$ for p. The best way to account for this from the modular viewpoint seems to involve the introduction of a subalgebra $\underline{u}_n$ of the "hyperalgebra" $U_K = U_{\mathbb{Z}} \otimes K$ (where $U_{\mathbb{Z}}$ is Kostant's $\mathbb{Z}$-form of the ordinary universal enveloping algebra), cf. Appendix U. This approach, suggested to the author by D.-N. Verma, has not been worked out fully and would at any rate complicate the notation at this point. It seems preferable to work mainly over the prime field for the time being.

The Weyl group W of G, and an associated affine Weyl group W_a (relative to p), exert enormous influence on the representation theory of G, Γ, $\underline{u}$. In particular, what Verma conjectured under the rubric "Harish-Chandra principle in prime characteristic" (to be called simply the "linkage principle (L)" here) plays a vital role in the modular theory. This principle, proved for large enough p by the author and (more recently) for all p by Kac and Weisfeiler, asserts that composition factors of an indecomposable module for G or $\underline{u}$ have "linked" highest weights.

One of our main techniques is tensoring with projective modules. Here the Steinberg module St (of dimension p^m) plays a central role, because it is simultaneously irreducible and projective for the algebras $K\Gamma$, $\underline{u}$. Moreover, St is obtained from an ordinary representation of Γ by reduction mod p. (In the context of character theory, it might be noted that the technique of multiplying by the Steinberg character was already used by Schur in his determination of the characters of SL(2,q).) A brief sketch of the various aspects of the Steinberg representation is given in Appendix S.

The paper is organized as follows: Part I reviews, for the sake of coherence, what is known about the irreducible modular representations of G, $\underline{u}$, Γ_n, along with the linkage principle. In Part II some basic results about PIM's are proved, including those announced in Humphreys, Verma [1]. Part III deals with the ordinary representations of Γ as they relate to the modular ones. Part IV discusses analogous results for the twisted groups of types A, D, E_6. Appendix R lists some standard facts from representation theory which are needed. Since notation and terminology vary quite a bit (especially in the author's own previous articles), a list of notation is provided for the reader's convenience. There is a fairly comprehensive bibliography.

This was originally planned to be a joint paper with D.-N. Verma, who is responsible for some of the main ideas in Part II and whose advice has been of great help. But factors of time and distance have conspired to defeat that intention. I am grateful to John W. Ballard for sending me a copy of his thesis, in which some of the results of Part II were discovered independently from the point of view of Brauer characters. Correspondence with Jens C. Jantzen has led to a more precise formulation of the ideas in Part III. Among those who have kindly supplied preprints of their work are R. W. Carter, S. G. Hulsurkar, J. C. Jantzen, A. V. Jeyakumar, G. Lusztig. It is a pleasure to acknowledge the hospitality of Helmut Behr and the Fakultät für Mathematik, Universität Bielefeld, who provided the occasion for some lectures on this subject matter during May-June 1974. Research was partially supported by National Science Foundation grants.

I. IRREDUCIBLE MODULAR REPRESENTATIONS

As references for the results stated without proof in this part, we suggest: Borel [1,2], Curtis [4], Humphreys [1,3,8], Jantzen [1], Steinberg [3,4,5], Verma [2], Wong [2].

§1. Weights and maximal vectors

1.1 Weights

Fix a maximal torus T of G, B = TU a Borel subgroup containing T, where U is the unipotent radical of B. Let Φ be the root system of G, Φ^+ (resp. Φ^-) the set of positive (resp. negative) roots relative to B, $\alpha_1,\ldots,\alpha_\ell$ the corresponding simple roots (ℓ = rank G), W the Weyl group, σ_α the reflection in W corresponding to $\alpha \in \Phi$, $\sigma_i = \sigma_{\alpha_i}$, σ_o the unique element of W sending Φ^+ to Φ^-. The cardinality of Φ^+ will be denoted by m, so dim G = ℓ + 2m.

Since G is simply connected, the group X = X(T) of rational characters of T is the full lattice of weights, with a basis consisting of the fundamental dominant weights $\lambda_1,\ldots,\lambda_\ell$, where $2(\lambda_i,\alpha_j)/(\alpha_j,\alpha_j)$ = δ_{ij} for any nondegenerate W-invariant symmetric bilinear form (λ,μ) on X. Set $<\lambda,\mu> = 2(\lambda,\mu)/(\mu,\mu)$. Define $\delta = \Sigma\lambda_i$ (equal to half the sum of positive roots). Denote by X^+ the set $\Sigma\mathbf{Z}^+\lambda_i$ of all dominant weights, by X_r the sublattice of X generated by the roots. W acts naturally on X and X_r, and acts trivially on the fundamental group X/X_r. There is a natural partial ordering of X : $\mu \leq \lambda$ if $\lambda-\mu$ is a sum of positive roots.

1.2 Maximal vectors

By G-module we shall always mean a finite dimensional rational G-module V. T acts completely reducibly on V, so V is the direct sum of weight spaces $\{v \in V | t.v = \lambda(t)v\}$ for various weights $\lambda \in X$. In case

$v \neq 0$ is a weight vector of weight λ which is fixed by all $u \in U$, we call v a maximal vector (of weight λ). Similarly, if v is fixed by all $u \in U^-$ (the unipotent part of the opposite Borel subgroup $\sigma_0 B \sigma_0$), we call v a minimal vector. The weight λ of a maximal vector is necessarily dominant and the weights μ occurring in the G-submodule generated by such a vector satisfy $\mu \leq \lambda$.

There is a natural basis $\{X_\alpha, Y_\alpha, \alpha \in \Phi^+; H_i, 1 \leq i \leq \ell\}$ of g, coming from a Chevalley basis of the semisimple Lie algebra $g_\mathbb{C}$ having root system Φ. When V is a G-module, the derived action of g on V makes V a restricted g-module, i.e., a $\underline{u}$-module, and each weight space of V is stable under $\underline{t}$ = Lie(T) or its restricted universal enveloping algebra $\underline{t}$. The differential of a weight $\lambda \in X$ is a linear function on $\underline{t}$ which we continue to write as λ and call a weight. However, the elements of pX have zero differential, so the distinct weights of $\underline{t}$ correspond 1-1 to the elements of $\Lambda = X/pX$, a set of cardinality p^ℓ which we call the set of restricted weights. Sometimes it is convenient to identify Λ with the subset X_p of X^+ consisting of all λ for which $0 \leq <\lambda, \alpha_i> < p$ $(1 \leq i \leq \ell)$.

In a $\underline{u}$-module V, a maximal vector is defined to be a nonzero weight vector (relative to $\underline{t}$) which is killed by the Lie algebra $\underline{n}^+$ of U, i.e., by all X_α. Evidently a maximal vector of a G-module V is also a maximal vector for $\underline{u}$ (of the "same" weight); but not conversely. The following fact is easy to verify.

LEMMA. Let V be a G-module, v a vector of weight λ relative to T. Then $X_\alpha \cdot v$, $Y_\alpha \cdot v$ are also weight vectors relative to T, of respective weights $\lambda + \alpha$, $\lambda - \alpha$.

1.3 Formal Characters

It is useful to attach a formal character ch(V) to a G-module V. Let $\mathbb{Z}[X]$ be the group ring of X, with basis consisting of symbols $e(\lambda)$ in 1-1 correspondence with the elements of X, and with multiplication determined by the rule $e(\lambda)e(\mu) = e(\lambda + \mu)$. Let $m_V(\lambda)$ be the multiplicity of λ as a weight of V (the dimension of the corresponding weight space), and set $ch(V) = \sum_{\lambda \in X} m_V(\lambda)e(\lambda) \in \mathbb{Z}[X]$. The sum is of course finite.

§2. Irreducible modules

In this section we review briefly the construction of irreducible modules for G, $\underline{u}$, $K\Gamma_n$, and indicate some unsolved problems.

2.1 The modules M_λ

A finite dimensional irreducible module for the Lie algebra $\underline{g}_{\mathbb{C}}$ is uniquely determined up to isomorphism by the highest weight involved (the weight of a maximal vector), which may be regarded as an element of X^+. Denote by V_λ an irreducible module of highest weight λ, and by $ch(\lambda)$ its formal character in $\mathbb{Z}[X]$. The dimension $\dim(\lambda)$ of V_λ, the weight multiplicities $m_\lambda(\mu)$, and the formal character $ch(\lambda)$, may be computed using the methods of Weyl, Freudenthal, Kostant, Demazure.

Choice of an admissible lattice in V_λ allows one to "reduce modulo p", thereby obtaining a $\underline{u}$-module, which is derived from a G-module since G is of universal type. Call $\bar{V}_\lambda$ the G-module which results from the use of a minimal admissible lattice. It is known that $\bar{V}_\lambda$ is generated (as a G-module) by a maximal vector of weight λ and is indecomposable. The unique composition factor involving this maximal vector is denoted M_λ. (Carter, Lusztig [1] call $\bar{V}_\lambda$ a "Weyl

module" and describe it in more detail when G = SL(n,K), or rather
GL(n,K).)

THEOREM (Chevalley, Kostant). The G-modules M_λ ($\lambda \in X^+$) are pairwise nonisomorphic and exhaust the isomorphism classes of irreducible
G-modules.

Write p-ch(λ) for the formal character and p-dim(λ) for the
dimension of M_λ.

If a different admissible lattice in V_λ had been chosen, the
resulting G-module might not be isomorphic to $\bar{V}_\lambda$ (though it seems
likely to be indecomposable in any event). However, the composition
factors (one of them being M_λ) are well determined, because the p-ch(λ)
($\lambda \in X^+$) form a basis of the W-invariants in $\mathbf{Z}[X]$, just as the ch(λ) do
(cf. Bourbaki [$\underline{1}$, VI, §3]).

THEOREM (Curtis). Those M_λ for which $\lambda \in \Lambda$ (identified with X_p)
remain irreducible as $\underline{u}$-modules and exhaust the distinct isomorphism
classes of irreducible $\underline{u}$-modules.

Next, each M_λ ($\lambda \in X_p$) yields further irreducible G-modules $M_\lambda^{(p^k)}$
if the entries of the representing matrices are all raised to the p^k
power. Given $\lambda \in X^+$, there is a unique decomposition
$\lambda = \mu_0 + p\mu_1 + \ldots + p^{n-1}\mu_{n-1}$ for some n, with $\mu_i \in X_p$. Set
$X_q = \{\Sigma c_i \lambda_i \mid 0 \leq c_i < q\}$, $q = p^n$.

THEOREM (Steinberg). Let $\lambda \in X_q$ be written as above. Then M_λ is
isomorphic as G-module to $M_{\mu_0} \otimes M_{\mu_1}^{(p)} \otimes \ldots \otimes M_{\mu_{n-1}}^{(p^{n-1})}$, a
"twisted tensor product". For a given n, these M_λ form a complete set
of distinct irreducible $K\Gamma_n$-modules.

(For SL(2,q), this goes back to Brauer, Nesbitt [1].)

We remark that it is also possible to construct M_λ as a G-sub-module of the ring K[G] of polynomial functions on G. There M_λ sits inside a possibly larger space of functions (cf. Humphreys [8, 31.4]) which is isomorphic to the dual of $\bar{V}_\lambda$, in view of recent work of Bai, Musili, Seshadri, Kempf; Jantzen [3, Satz 1] uses this to show that $\bar{V}_\lambda$ has intrinsic meaning for G, as the "universal" G-module of highest weight λ.

2.2 The Steinberg module

The G-module $St = M_{(p-1)\delta}$, called the Steinberg module, plays a prominent role in all that follows; similarly for $St_n = M_{(p^n-1)\delta}$. (See Appendix S.)

THEOREM (Steinberg). $St = \bar{V}_{(p-1)\delta}$, so its dimension is p^m. All other M_λ ($\lambda \in X_p$) have strictly smaller dimension. Similarly, $\dim St_n = (p^n)^m$.

Proof. For the first assertion, see Steinberg [3, 8.2], based on the existence of a representation of Γ over $\mathbb{C}$ of degree p^m constructed in Steinberg [2]. (Another approach, based on the linkage principle, will be discussed in (4.1).) The second assertion follows easily from Weyl's dimension formula, while the third follows easily from the twisted tensor product theorem in (2.1).

2.3 Weight multiplicities

The main unsolved problem concerning the modules M_λ is the determination of their formal characters (or weight multiplicities) and dimensions. In principle, $p\text{-ch}(\lambda) = \Sigma a_{\lambda\mu} \text{ch}(\mu)$ for some integers $a_{\lambda\mu}$, where we can further specify that $a_{\lambda\lambda} = 1$ and that $a_{\lambda\mu} = 0$ unless $\mu \leq \lambda$. In other words, $M_\lambda \underset{G}{\leftrightarrow} \Sigma a_{\lambda\mu} \bar{V}_\mu$. Similarly

$ch(\lambda) = \Sigma b_{\lambda\mu} p\text{-}ch(\mu)$, with $b_{\lambda\mu} \in Z^+$; this records the composition factors of $\bar{V}_\lambda$. Thanks to Steinberg's twisted tensor product theorem, it would be enough to obtain this kind of information about the collection $\{M_\lambda | \lambda \in X_p\}$. However, it must be pointed out that even if $\lambda \in X_p$, some of the dominant weights below λ in the partial ordering may lie outside X_p unless Φ has type A_1, A_2, B_2 (cf. Verma [2,§4]).

For a given λ and a given p, it is possible (at least in principle) to compute effectively the weight multiplicities of M_λ. Burgoyne [1] has carried out computer calculations along this line for small ranks and small p. The underlying idea is to write down a square matrix of integers (of size equal to the dimension of a weight space of V_λ); the number of elementary divisors divisible by p counts the decrease in the dimension of this weight space when we pass to M_λ. Wong [1,2] and Jantzen [1,2,3] have exploited this idea to obtain some theoretical results; in particular, Jantzen gets good information about the determinant of the integral matrix involved, which in some cases is enough to solve the multiplicity problem completely. (See §4 below.)

2.4 Dual modules

The following fact is well known:

PROPOSITION. Let $\lambda \in X^+$. Then the dual G-module M_λ^* is isomorphic to $M_{-\sigma_o \lambda}$.

For example, St is isomorphic to its dual.

§3. The linkage principle

3.1 Indecomposable modules

Unlike the situation in characteristic 0, G-modules need not be completely reducible. So the representation theory of G involves not just the irreducible modules M_λ, but also other indecomposable modules. Among these are the modules $\bar{V}_\lambda$ ($\lambda \in X^+$), which are known to be indecomposable also for $\underline{u}$ when $\lambda \in \Lambda$. Other indecomposable G-modules will play a significant role in Part II; but it is fair to say that not much is known about indecomposable G-modules in general.

As mentioned in (2.3), a basic problem is to decide when $\bar{V}_\lambda$ is the same as M_λ. In his proof of Weyl's character formula for algebraic groups (in characteristic 0), Springer [1] obtained as a by-product a rather weak sufficient condition for equality to hold. But slightly earlier Verma had formulated a much more precise condition on highest weights of composition factors of an indecomposable module for G (or $\underline{u}$), called by him the "Harish-Chandra principle in prime characteristic". This conjectured condition has only recently been proved in full generality by Kac, Weisfeiler [1].

Define two weights $\lambda, \mu \in X$ to be linked (written $\lambda \sim \mu$) if there exists $\sigma \in W$ for which $\sigma(\lambda + \delta) \equiv \mu + \delta \pmod{pX}$. This is an equivalence relation on X. Otherwise formulated, λ and μ are linked if they are conjugate under the transformation group $\tilde{W}$ on X generated by W along with all translations by elements of pX, where $\sigma \in \tilde{W}$ acts by the rule: $\sigma \cdot \lambda = \sigma(\lambda + \delta) - \delta$. We say that two weights $\lambda, \mu \in \Lambda$ are linked if $\sigma(\lambda + \delta) = \mu + \delta$ in Λ for some $\sigma \in W$.

In this language, Verma's conjecture can be formulated as follows:

LINKAGE PRINCIPLE (L). If M_λ, M_μ are composition factors of an indecomposable G-module (resp. $\underline{u}$-module, with $\lambda, \mu \in \Lambda$), then λ and μ are linked in X (resp. in Λ).

The conjecture was motivated by a classical theorem of Harish-Chandra (cf. Humphreys [3, §23]), which defines an isomorphism of the center of the universal enveloping algebra of g_C onto the algebra of W-invariants in the universal enveloping algebra of a Cartan subalgebra (this latter algebra being a polynomial algebra on ℓ generators). This isomorphism allows one to read off the eigenvalue with which any element of the center (e.g., the Casimir element) acts on a g_C-module (not necessarily finite dimensional) generated by a maximal vector of weight λ (not necessarily dominant). If λ and μ are weights which give rise to the same eigenvalue for every element of the center, it follows easily that $\sigma(\lambda + \delta) = \mu + \delta$ for some $\sigma \in W$.

We remark that an obvious necessary condition for M_λ and M_μ to occur as composition factors of an indecomposable G-module is that λ and μ lie in the same coset of X/X_r, in view of the way root subgroups of G act on weight vectors. But no such constraint exists for $\underline{u}$-modules, as the modules Z_λ of §5 will show.

3.2 Verification of (L)

THEOREM. (L) is true.

The proof was carried out initially under the added assumption that p exceeds the Coxeter number of W (in Humphreys [1]). Here the idea was simply to imitate the proof of Harish-Chandra's theorem, with $\underline{u}$ in place of the universal enveloping algebra. This method is fairly constructive and yields a homomorphism (perhaps not injective) from the center of $\underline{u}$ onto the algebra of W-invariants in the restricted enveloping algebra $\underline{h}$. This easily implies the linkage principle for indecomposable $\underline{u}$-modules. The transition to indecomposable G-modules is not too difficult. However, the entire method breaks down badly for small p.

In their study of type A_ℓ (where they looked at general linear rather than special linear groups), Carter and Lusztig [1] found explicit generators over $\mathbb{Z}$ for the center of the Kostant $\mathbb{Z}$-form of the universal enveloping algebra. Reducing mod p, they could read off the eigenvalues of the resulting central elements on the various modules $\bar{V}_\lambda$. Again the linkage principle follows easily, although they stated the conclusion only in connection with the question of when there could exist a non-zero G-module homomorphism $\bar{V}_\lambda \rightarrow \bar{V}_\mu$. No condition is placed on p.

Quite recently Kac and Weisfeiler [1] have written down a general proof of (L). They study the center of the universal enveloping algebra of $\underline{g}$ (not just the center of $\underline{u}$), using older results of Zassenhaus, and show by nonconstructive methods that the analogue of Harish-Chandra's homomorphism is surjective; in fact, it induces an isomorphism from the G-invariants in the center onto the W-invariants in the universal enveloping algebra of $\underline{h}$.

3.3 The affine Weyl group

Verma [2, §5] has reformulated the notion of linkage in terms of an affine Weyl group W_a (his $\tilde{W}'$), the subgroup of $\tilde{W}$ generated by W along with all translations by elements of pX_r. We noted above that if M_λ, M_μ are composition factors of an indecomposable G-module, then necessarily $\lambda - \mu \in X_r$; (L) further implies that λ, μ must be $\tilde{W}$-conjugate. Verma observes that when p does not divide $f = [X : X_r]$, these two conditions imply (hence are equivalent to) the condition that λ, μ are W_a-conjugate, in which case they may be said to be W_a-linked.

Suppose that p does divide f. It is still reasonable to conjecture that the highest weights of composition factors in an indecomposable G-module must be W_a-linked. The argument of Carter, Lusztig [1] leads to this conclusion in type A_ℓ, while Jantzen [3] verifies it for some small ranks, including the cases B_2, B_3, C_4, D_4, in the case of composition factors of modules $\bar{V}_\lambda$.

3.4 Some examples

The geometry of affine reflection groups such as W_a has been well studied (cf. Bourbaki [1], Verma [2]). The usual affine Weyl group associated with Φ is generated by $\sigma_1, \ldots, \sigma_\ell$ along with the reflection relative to the highest short root of Φ, while for W_a we multiply this root by p and translate everything by $-\delta$ (leaving the geometry essentially unaffected). The complement in $X \otimes \mathbb{R}$ of the union of reflecting hyperplanes is partitioned into disjoint alcoves (each a euclidean ℓ-simplex), the closure of any one being a fundamental domain for the action of W_a. X_p lies in the union of a number of alcoves ($|W|/f$, to be precise), the "top" one containing $(p-1)\delta$ and the "lowest" one containing $-\delta$. More exactly, the alcoves meeting X_p contain precisely those $\lambda = \Sigma c_i \lambda_i$ for which $-1 \leq c_i \leq p-1$. Call a weight p-regular if it lies in the interior of an alcove.

A couple of illustrations (cf. Verma [2, §3]) will be helpful. For type A_1, X_p lies in a single alcove; here we always identify weights $r\lambda_1$ with integers r:

A_1

$$\begin{array}{ccc} -1 & 0 & \quad\quad\quad p-1 \end{array}$$

For type A_2 (resp. B_2) there are two (resp. four) alcoves:

A_2

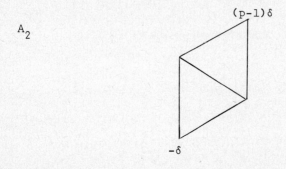

$(p-1)\delta$

$-\delta$

B_2

$(p-1)\delta$

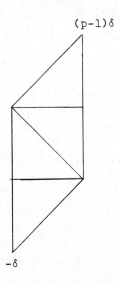

$-\delta$

For G_2 or for higher ranks, it is convenient to depict the relationship among alcoves schematically, ordering them from top to bottom
according to their shared walls. G_2 has 12 alcoves meeting X_p, while
A_3 has 6:

G_2

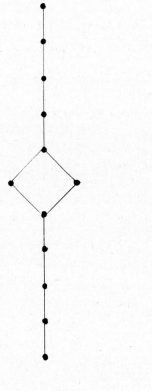

A_3

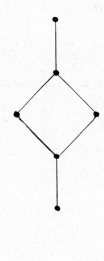

§4. Application to M_λ

In this section we draw some inferences from the linkage principle, and summarize some recent work on the modules M_λ.

4.1 Irreducibility criterion

Recall the equation $p\text{-ch}(\lambda) = \Sigma a_{\lambda\mu}\text{ch}(\mu)$, summed over $\mu \in X^+$; here $a_{\lambda\lambda} = 1$ and $a_{\lambda\mu} \neq 0$ implies $\mu \leq \lambda$. Thanks to (L), we can add the further condition: $\mu \sim \lambda$.

THEOREM (cf. Verma [2, 5.2]). Let $\lambda \in X^+$. If no dominant weight $\mu < \lambda$ satisfies $\mu \sim \lambda$, e.g., if λ lies in the closure of the lowest alcove of X_p, then $\bar{V}_\lambda = M_\lambda$.

COROLLARY. If $\lambda = (p-1)\delta$, then $\bar{V}_\lambda = M_\lambda$ (= St). In particular, $\dim \text{St} = p^m$ and $\dim \text{St}_n = (p^n)^m$.

The corollary follows, because $(p-1)\delta$ is minimal among the dominant weights in its linkage class in X. The corollary was stated in (2.2) without the use of (L), but once (L) is known the present method of proof is more direct.

For a p-regular weight λ, the converse of the theorem ought to be true: $\bar{V}_\lambda = M_\lambda$ implies that λ lies in the lowest alcove of X_p. This is proved in Carter, Lusztig [1, p. 232] when G is of type A_ℓ (their proof for the general linear group adapts at once to the special linear group).

4.2 Composition factors of $\bar{V}_\lambda$

One wants to know in general how to find the integers $a_{\lambda\mu}$ for which $p\text{-ch}(\lambda) = \Sigma a_{\lambda\mu}\text{ch}(\mu)$, or equivalently, how to find the non-negative integers $b_{\lambda\mu}$ for which $\text{ch}(\lambda) = \Sigma b_{\lambda\mu}p\text{-ch}(\mu)$.

Verma [2, Conjecture II] has proposed a further necessary condition for $a_{\lambda\mu}$ to be nonzero, beyond the requirements that $\mu \leq \lambda$ and that μ be linked to λ. This further condition is, roughly, that μ be obtainable from λ by applying successive reflections in W_a, each reflected weight being lower in the partial order. Jantzen [2, Theorem 3] has proved Verma's conjecture for G of type A_ℓ; in Jantzen [3, Theorem 2] he extends this to most types, under mild restrictions on p.

Beyond this, Verma [2, Conjecture III] (and, independently, the author) has suggested that the value of $a_{\lambda\mu}$ should depend just on the relative positions of the alcoves to which λ and μ belong, assuming that λ,μ are p-regular. This conjecture has been proved by Jantzen [2, Theorem 1] (under the hypothesis that (L) holds), who also obtains some information about the $a_{\lambda\mu}$ in the irregular cases.

Verma [2, Conjecture V] has also conjectured that the $a_{\lambda\mu}$ should be the same for all p, depending only on the relative positions of the alcoves involved. This has been observed to be true in a number of cases, but Jantzen has recently pointed out exceptions for certain small p. It seems likely, nevertheless, that the conjecture will hold for all but small p.

The known values of the $a_{\lambda\mu}$ do not look too "wild", but neither do they seem to fit into a simple, predictable pattern. They alternate in sign as we pass from one alcove across a wall into a neighboring alcove, which is partly explained by Jantzen's method but which cannot yet be predicted in general. The values of the $a_{\lambda\mu}$ seem to be ±1 until a branching of alcoves is encountered, which might also turn out to be a general pattern.

4.3 Some special cases

Let us survey some of the known results.

A_1: When $G = SL(2,K)$, (4.1) implies that all $\bar{V}_\lambda = M_\lambda$ ($\lambda \in X_p$);
this is of course easy to verify directly, cf. Brauer, Nesbitt [1].
From the twisted tensor product theorem one then gets detailed
information about all M_λ. Recently, Carter and Cline [1] have been
able to describe explicitly the submodule structure of $\bar{V}_\lambda$ ($\lambda \in X^+$).

A_2: When $G = SL(3,K)$, Braden [1] obtained the u-composition
factors of $\bar{V}_\lambda$ for $\lambda \in \Lambda$ (omitting a few small p). He found in effect
that for λ in the top alcove, $\bar{V}_\lambda \overset{\leftrightarrow}{\underset{u}{}} M_\lambda + M_{\lambda^o}$, where λ^o is the linked
weight in the bottom alcove lying below λ in the partial order, i.e.,
$\lambda^o = \sigma_o(\lambda + \delta) - \delta$ in Λ (= the weight gotten by reflecting λ across the
wall dividing the two alcoves). On the other hand, for λ in the
closure of the bottom alcove, $\bar{V}_\lambda = M_\lambda$ (cf. (4.1)). This recipe also
describes the G-composition factors, since all highest weights
involved are restricted.

B_2: Braden obtained partial results, which have recently been
completed by Jantzen [3]: Number the alcoves in X_p from top to
bottom as 1, 2, 3, 4. If λ^1 lies in the (interior of) the top alcove,
let λ^2, λ^3, λ^4 denote the linked weights below λ^1 in the partial
order. Then p-ch(λ^1) = ch(λ^1) - ch(λ^2) + ch(λ^3) - ch(λ^4), p-ch(λ^2) =
= ch(λ^2) - ch(λ^3) + ch(λ^4), p-ch($\lambda 3$) = ch(λ^3) - ch(λ^4), p-ch(λ^4) = ch(λ^4).
When a weight λ lies in a wall of an alcove, we usually get p-ch(λ) =
= ch(λ), with the exceptions pictured below (the second occurring just
when p = 2).

A_3: For $G = SL(4,K)$, Jantzen [2] gives explicit descriptions of
the p-ch(λ) for $\lambda \in X_p$ in terms of the various ch(μ), where $\mu \in X^+$ and
$\mu \leq \lambda$. This is complicated somewhat by the occurrence of nonrestricted
μ: there are two nonrestricted alcoves below the top one, as
indicated below by dotted lines. Number the alcoves as shown.
Jantzen finds formulas comparable to those for B_2, with alternation
of sign whenever a wall is crossed:

B_2

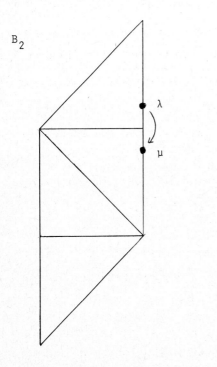

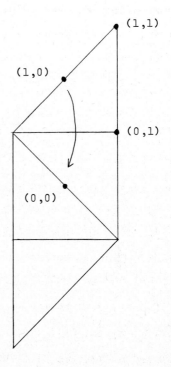

$p - ch(\lambda) = ch(\lambda) - ch(\mu)$

$2 - ch(1,0) = ch(1,0) - ch(0,0)$

A_3

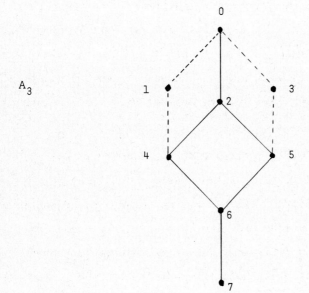

$$p\text{-}ch(\lambda^0) = ch(\lambda^0) - ch(\lambda^1) - ch(\lambda^2) - ch(\lambda^3)$$
$$+ ch(\lambda^4) + ch(\lambda^5) - 2ch(\lambda^6) + 3ch(\lambda^7)$$

$$p\text{-}ch(\lambda^1) = ch(\lambda^1) - ch(\lambda^4) + ch(\lambda^6) - ch(\lambda^7)$$

$$p\text{-}ch(\lambda^2) = ch(\lambda^2) - ch(\lambda^4) - ch(\lambda^5) + ch(\lambda^6) - 2ch(\lambda^7)$$

$$p\text{-}ch(\lambda^3) = ch(\lambda^3) - ch(\lambda^5) + ch(\lambda^6) - ch(\lambda^7)$$

$$p\text{-}ch(\lambda^4) = ch(\lambda^4) - ch(\lambda^6) + ch(\lambda^7)$$

$$p\text{-}ch(\lambda^5) = ch(\lambda^5) - ch(\lambda^6) + ch(\lambda^7)$$

$$p\text{-}ch(\lambda^6) = ch(\lambda^6) - ch(\lambda^7)$$

$$p\text{-}ch(\lambda^7) = ch(\lambda^7).$$

G_2: For $p = 3$, see Springer [1]. When $p \geq 7$, Jantzen [3] has calculated the numbers $a_{\lambda\mu}$ (or $b_{\lambda\mu}$). Here again, not all $a_{\lambda\mu}$ are ± 1. (The case $p = 5$ is not settled.)

A_ℓ: Carter, Lusztig [1] have obtained some nonzero G-module homomorphisms $\bar{V}_\mu \to \bar{V}_\lambda$, by making detailed calculations in the Kostant Z-form of the universal enveloping algebra and reducing mod p. Recently Carter's research student M. T. J. Payne has been able to show that $\text{Hom}(\bar{V}_\mu, \bar{V}_\lambda)$ has dimension precisely 1 in many cases; Jantzen [3] also has obtained such results in the general setting.

It should be added that Jantzen [4] has found some generic decomposition patterns for $\bar{V}_\lambda$ when λ is not restricted. These will be discussed further in Part III.

II. PROJECTIVE MODULES

Now we turn to the study of projective modules, first for $\underline{u}$ and then for $K\Gamma$. Recall (R.1) that in either case, a projective module can be written in an essentially unique way as a direct sum of principal indecomposable modules (PIM's), the latter being in natural 1-1 correspondence with the irreducible modules M_λ ($\lambda \epsilon \Lambda$). We denote by Q_λ (resp. R_λ) the PIM of $\underline{u}$ (resp. $K\Gamma$) having M_λ as its unique top (and bottom) composition factor. The main result of this part (cf. (8.2), (10.2)) asserts that Q_λ is in a natural way a G-module, whose restriction to Γ is a projective $K\Gamma$-module involving R_λ as a direct summand. This is analogous to the relationship between M_λ as $\underline{u}$-module and M_λ as $K\Gamma$-module; but dim Q_λ is sometimes larger than dim R_λ.

§5. The $\underline{u}$-modules Z_λ and Q_λ

5.1 The modules Z_λ

The analogue for $\underline{u}$ of a Verma module is constructed in Humphreys [1] as an aid in the proof of the linkage principle. Here we recall briefly the main features of that construction.

Each $\lambda \epsilon \Lambda$ defines in an obvious way a restricted representation of the subalgebra $\underline{b}$ of $\underline{g}$, having degree 1, i.e., a $\underline{b}$-module M spanned by a single vector m. The induced $\underline{u}$-module $\underline{u} \otimes_{\underline{b}} M$ is then called Z_λ. It has a basis consisting of the p^m vectors $Y_{\alpha_1}^{i_1} \ldots Y_{\alpha_m}^{i_m}(1 \otimes m)$, $0 \le i_k < p$, where the positive roots $\alpha_1, \ldots, \alpha_m$ are given in any fixed order. Moreover, Z_λ is generated as $\underline{u}$-module by the maximal vector $1 \otimes m$ of weight λ. This suggests an alternative description by generators and relations: Z_λ is isomorphic to the quotient of $\underline{u}$ by its left ideal generated by all X_α and all $H_i - \lambda(H_i) \cdot 1$ $(1 \le i \le \ell)$, with $1 \otimes m$ corresponding to the coset of 1. It follows that Z_λ has the characteristic universal mapping property of a Verma module (cf. Dixmier [1, Ch. 7], Verma [1]): Any $\underline{u}$-module generated by a maximal vector of weight λ (for example, M_λ or $\bar{V}_\lambda$) is a homomorphic image of Z_λ. Some facts proved in the cited paper are listed in the following theorem.

THEOREM. Let $\lambda, \mu \in \Lambda$.

(a) Up to scalar multiples, Z_λ has a unique minimal vector, corresponding to the coset of $Y_{\alpha_1}^{p-1} \ldots Y_{\alpha_m}^{p-1}$ $(\alpha_1, \ldots, \alpha_m$ any ordering of the positive roots), whose weight is $\lambda - 2(p-1)\delta$.

(b) Z_λ has a unique minimal submodule, isomorphic to $M_{\sigma_o(\lambda+\delta) - \delta}$ $(\sigma_o$ = longest element of W). In particular, Z_λ is indecomposable.

(c) If $f : Z_\lambda \to N$ is an epimorphism of $\underline{u}$-modules, the criterion for f to be an isomorphism is that $Y_{\alpha_1}^{p-1} \ldots Y_{\alpha_m}^{p-1}$ not annihilate the image of $1 \otimes m$ in N.

(d) If $\lambda = (p-1)\delta$, then $Z_\lambda \cong M_\lambda$ (= St); but otherwise M_λ is a proper homomorphic image of Z_λ.

(e) If $\lambda \sim \mu$, then $Z_\lambda \overset{\leftrightarrow}{\underline{u}} Z_\mu$.

A few comments are in order. The proof of (a) is elementary, and from it (b) follows quickly, since the lowest weight $\lambda - 2(p-1)\delta$ accompanies the highest weight $\sigma_o(\lambda - 2(p-1)\delta) = \sigma_o\lambda + 2(p-1)\delta =$ $= \sigma_o(\lambda + \delta) - \delta$ (equality as elements of Λ). From (a) and (b) we immediately get (c). Since dim $Z_\lambda = p^m$, (d) follows from (2.2). The proof of (e) is rather easy and will be recalled in (5.3) below. (None of this presupposes the linkage principle.)

The weight $\sigma_o(\lambda + \delta) - \delta$ which occurs in (b) will be called the linked weight <u>opposite</u> λ and written λ^o.

5.2   Decomposition numbers

From parts (b), (e) of Theorem 5.1, along with the linkage principle, it follows that the composition factors of Z_λ are precisely those M_μ for which $\mu \sim \lambda$, each taken with a multiplicity $d_\mu > 0$ which depends only on μ. These decomposition numbers d_μ turn out to be of significant interest. It seems <u>a priori</u> that they might vary with the choice of p, with the choice of a linkage class, etc. But there is much evidence to suggest that they depend essentially just on which alcove of X_p a weight belongs to. For example, we shall see in (9.2) that $d_\mu = 1$ when μ is a p-regular weight maximal in its linkage class.

The numbers d_μ can be effectively computed in certain cases by combining a knowledge of the numbers $a_{\lambda\mu}$ (2.3) with some results of Verma and Hulsurkar to be described in §6. For type A_1 the computation is rather trivial, while for A_2 it is easy (a fact not perceived by the author in Humphreys [1], but pointed out by Verma [2]): On the one hand, there is a surjection $Z_\lambda \rightarrow \bar{V}_\lambda$, so that Z_λ involves the composition factors of $\bar{V}_\lambda$ at least once. When λ is in the top alcove, λ^o in the bottom alcove, the (nonzero) kernel of this surjection contains the unique minimal submodule M_{λ^o} of Z_λ (cf. (5.1)), which is the same as $\bar{V}_{\lambda^o}$. So in fact Z_λ includes among its composition factors those of the direct sum of all $\bar{V}_\mu$ (μ linked to λ). But an easy calculation using Weyl's dimension formula shows that the latter direct sum already has dimension p^m. For λ in the top alcove (with $p \neq 3$), it follows that $d_\lambda = 1$ and $d_{\lambda^o} = 2$.

In the examples below, the partial ordering of alcoves covering X_p is indicated schematically, with the value of d_μ for all (p-regular) μ in an alcove written alongside, it being assumed that p does not

24

divide f. For B_2, G_2, A_3, the values of $a_{\lambda\mu}$ are based on calculations of Jantzen [2,3], cf. (4.3).

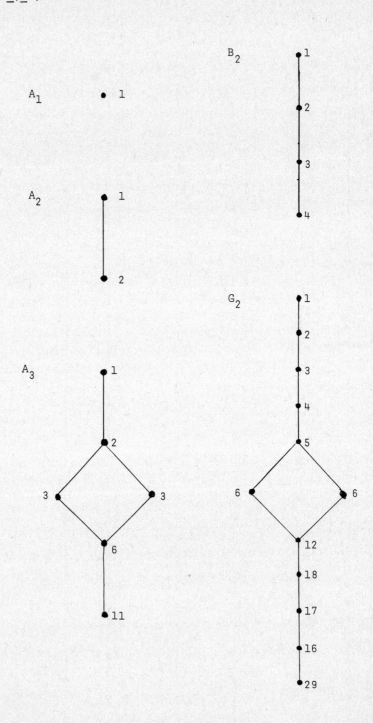

The tendency of the d_μ to increase monotonically from the top alcove downward until a branching occurs has not yet been adequately explained, but it contributes to the author's belief that these numbers may have intrinsic geometric significance for the Weyl group (or affine Weyl group). The lowest alcove for A_3 and the four lowest alcoves for G_2 yield peculiar values of d_μ, which result at least in part from the occurrence of certain nonrestricted weights in associated G-modules.

Notice that the sum of all d_μ, for μ running over the linkage class of a p-regular weight λ (i.e., the number of composition factors of Z_λ), is given by:

$$
\begin{aligned}
A_1 &: \quad 2 \\
A_2 &: \quad 9 \\
A_3 &: \ 104 \\
B_2 &: \ 20 \\
G_2 &: \ 119
\end{aligned}
$$

It appears (cf. Part III) that this is equal to the number of composition factors obtained when a "typical" irreducible $\mathbb{CT}$-module is reduced modulo p, the highest weights of these composition factors being distributed among alcoves just as for Z_λ. (This number also shows up in Jantzen [4], where it is called N(R), R being the root system in question.)

The reasoning which leads to the above examples in low ranks also suggests how to analyze the value of d_λ when λ fails to be p-regular. Assuming that decomposition numbers have already been assigned to alcoves, we should assign to a weight λ lying in a wall common to two alcoves the decomposition number belonging to the lower alcove (in the partial ordering of alcoves). More precisely, λ lies in the "upper closure" of a unique alcove (cf. Jantzen [2]) for this notion), and we should let d_λ be the decomposition number of this alcove. The

26

only exceptions to these statements should arise when p divides f, in
which case "ramification" can occur; see Verma [2, §7] for a discussion
of this phenomenon. Here we shall be content to give a few illustra-
tive examples, which have been done by ad hoc methods (cf. the tables
in Humphreys [5]).

For type A_2, p = 5, d_λ = 2 when λ is on or below the diagonal of
the parallelogram; otherwise d_λ = 1.

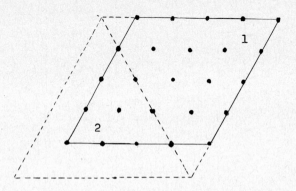

On the other hand, ramification occurs for type A_2, p = 3, yielding
$d_{(1,1)}$ = 3 and $d_{(0,0)}$ = 6 (3 times the usual values).

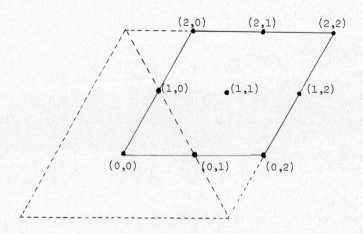

Finally, in type B_2, $p = 3$, precisely the expected values occur, e.g., $d_{(0,0)} = 4$.

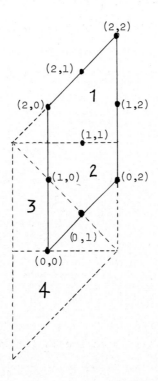

5.3 Intertwining operators

We digress briefly to discuss further the structure of the $\underline{u}$-modules Z_λ ($\lambda \in \Lambda$).

The proof of (L) in Humphreys [1] involves a study of $\underline{u}$-module homomorphisms $Z_\mu \to Z_\lambda$, some of which are quite easy to construct. Fix $\lambda \in \Lambda$. For each simple root α_i, set $n_i = \langle \lambda, \alpha_i \rangle$. Then $\mu = \lambda - (n_i+1)\alpha_i$ is linked to λ in Λ. If v^+ is a maximal vector generating Z_λ, it is not hard to see that $Y_i^{n_i+1} \cdot v^+$ is a maximal vector of weight μ (or else is 0 when $n_i + 1 = p$, $\mu = \lambda$), where we denote Y_{α_i} by Y_i. This determines a $\underline{u}$-module homomorphism $f : Z_\mu \to Z_\lambda$. The same procedure taken in reverse yields a homomorphism $g : Z_\lambda \to Z_\mu$, because

$\lambda = \mu - (<\mu, \alpha_i> + 1)\alpha_i$. The composite $g \circ f = 0$, in view of the fact that $<\mu, \alpha_i> = -n_i - 2$, forcing $Y_i^{<\mu,\alpha_i>+1} \ Y_i^{<\lambda,\alpha_i>+1} = Y_i^p = 0$. Similarly, $f \circ g = 0$.

Theorem 5.1(e) is proved by treating first the case just sketched, in which λ and μ are linked by a simple reflection σ_i: Ker f is readily seen to coincide with Im g, and similarly Ker g = Im f. Now the composition factors of Z_λ are those of Im f along with those of $Z_\lambda/\text{Im } f = Z_\lambda/\text{Ker } g \cong \text{Im } g$; these are also the composition factors of Z_μ. Transitivity of linkage then completes the proof.

It may be instructive to study $Z = \coprod_{\lambda \in \Lambda} Z_\lambda$, which can be viewed as the $\underline{u}$-module induced from the trivial 1-dimensional module for $\underline{n}^+$. For each simple root α_i and each pair λ, μ as above, we have just defined a map $Z_\mu \to Z_\lambda$ and a map $Z_\lambda \to Z_\mu$. These maps may all be combined into a single $\underline{u}$-module endomorphism $T_i : Z \to Z$, satisfying (*) $T_i^2 = 0$. Evidently T_i stabilizes the sum of all Z_λ as λ runs over a linkage class. For example, T_i is just the zero map of St into itself.

We assert that the maps T_i $(1 \le i \le \ell)$ also satisfy the condition: Given two reduced expressions $\sigma_{i(1)} \cdots \sigma_{i(t)} = \sigma_{j(1)} \cdots \sigma_{j(t)}$ in W, we have $T_{i(1)} \cdots T_{i(t)} = T_{j(1)} \cdots T_{j(t)}$. Because W is a Coxeter group, it suffices to verify this in rank 2, cf. Bourbaki [$\underline{1}$, IV, §1, Prop. 5], where we can apply the following identities of Verma [$\underline{1}$, (5)(6)(7)] in the universal enveloping algebra of $\underline{g}_{\mathbb{C}}$:

$(A_2) \quad Y_1^r Y_2^{r+s} Y_1^s = Y_1^s Y_2^{r+s} Y_1^r$

$(B_2) \quad Y_1^r Y_2^{r+s} Y_1^{r+2s} Y_2^s = Y_2^s Y_1^{r+2s} Y_2^{r+s} Y_1^r$

$(G_2) \quad Y_1^r Y_2^{r+s} Y_1^{2r+3s} Y_2^{r+2s} Y_1^{r+3s} Y_2^s =$
$\qquad\qquad Y_2^s Y_1^{r+3s} Y_2^{r+2s} Y_1^{2r+3s} Y_2^{r+s} Y_1^r$.

Here $r, s \in \mathbb{Z}^+$. These identities carry over at once to $\underline{u}$, where both sides become 0 if an exponent exceeds p-1, since we are using a

Chevalley basis. Now if $\sigma_1\sigma_2$ has order k in W (of rank 2), these identities are precisely what we need to establish:

(**) $\underbrace{T_1T_2T_1\cdots}_{\text{k factors}}$ = $\underbrace{T_2T_1T_2\cdots}_{\text{k factors}}$

Carter and Lusztig [2] have defined analogous operators inter-twining the representations of KΓ induced from 1-dimensional represen-tations of a Borel subgroup. They obtain relations on their T_i similar to (*) and (**), and show that these generate all relations among their T_i. For SL(3,p) they find typically 9 composition factors for an induced module (the number we typically find for Z_λ) and describe in detail the submodule structure via their T_i. It is tempting to conjecture that the algebra of endomorphisms of Z genera-ted by our T_i is defined completely by the relations of types (*) and (**).

5.4   The modules Q_λ

The modules Z_λ play an important role in the description of the PIM's Q_λ of $\underline{u}$. In view of (L), all composition factors of Q_λ have linked highest weights, allowing us to write:

$$Q_\lambda \underset{\underline{u}}{\leftrightarrow} \underset{\mu\sim\lambda}{\Sigma} c_{\lambda\mu}M_\mu.$$

The integers $c_{\lambda\mu}$ (called the Cartan invariants of $\underline{u}$) form a square matrix C of block diagonal form, one block for each linkage class in Λ. Since $\underline{u}$ is a "symmetric" algebra (R.3), C is in fact a symmetric matrix. The following theorem is proved in Humphreys [1,4.4,4.5]:

THEOREM. $Q_\lambda \underset{\underline{u}}{\leftrightarrow} \underset{\mu\sim\lambda}{\Sigma} d_\lambda Z_\mu \underset{\underline{u}}{\leftrightarrow} a_\lambda d_\lambda Z_\lambda \underset{\underline{u}}{\leftrightarrow} a_\lambda d_\lambda \underset{\mu\sim\lambda}{\Sigma} d_\mu M_\mu,$ where d_λ defined in (5.2), a_λ = cardinality of linkage class of λ. In parti-cular, dim $Q_\lambda = a_\lambda d_\lambda p^m$.

COROLLARY. If $\lambda = (p-1)\delta$, then $Q_\lambda = Z_\lambda = M_\lambda$ (= St), so the Steinberg module is projective. (A direct proof that St is projective will be given in (5.5), not presupposing (L).)

For $G = SL(2,K)$, $\underline{g} = \underline{sl}(2,K)$, the theorem implies that $\dim Q_\lambda = 2p$ unless $\lambda = p-1$. For $G = SL(3,K)$, $\dim Q_\lambda$ is typically either $6p^3$ or $12p^3$.

The proof of the theorem is rather lengthy, so we shall be content to sketch briefly the steps involved (cf. also Verma [2]). The first, and most complicated, step is to show that $Q_\lambda \leftrightarrow e_\lambda Z_\lambda$ for some integer e_λ. In turn, since $Z_\lambda \leftrightarrow \Sigma d_\mu M_\mu$ (sum over linkage class of λ), we have $c_{\lambda\mu} = e_\lambda d_\mu$. But $\underline{u}$ is a symmetric algebra (R.3), so $f_\lambda = e_\lambda/d_\lambda$ is a constant depending only on the linkage class. Now we consider the dimension of the block of $\underline{u}$ associated with λ. This can be calculated in two ways. Viewing $\underline{u}$ as a free $\underline{b}^-$ module, Z_μ ($\mu \sim \lambda$) will be a PIM of $\underline{b}^-$ and will occur a total of p^m times, all occurences being in this block, so the block has dimension $(a_\lambda p^m)p^m$. On the other hand, by general principles, the block involves $m_\mu = \dim M_\mu$ copies of Q_λ for each μ linked to λ. Since $\dim Q_\mu = e_\mu p^m$, the block dimension must be $p^m \Sigma m_\mu e_\mu$ (sum over linkage class of λ). Comparison of these two results shows that $f_\lambda = a_\lambda$, and from this the theorem follows.

One consequence of the proof of the first step of the theorem will be vital in §8:

COROLLARY (OF PROOF). Q_λ has a (nonunique) submodule isomorphic to Z_{λ_0}.

This statement does not require (L) at all for its proof. Notice that it is compatible with the fact that Q_λ has a unique minimal submodule M_λ, cf. Theorem 5.1(b).

5.5 Proof that St is projective

As promised in (5.4), we offer here a direct proof that St is a projective $\underline{u}$-module, without appealing to (L).

Let $f : M \to St$ be an epimorphism of $\underline{u}$-modules. Fix a maximal vector v^+ (resp. a minimal vector v^-) in St; these are unique up to scalar multiples. Choose a (nonunique) preimage m^+ of v^+ in M. Since $\underline{t}$ acts completely reducibly, it may be assumed that m^+ has weight $(p-1)\delta$ relative to $\underline{t}$; but m^+ might not be a maximal vector. Applying $Y_{\alpha_1}^{p-1} \ldots Y_{\alpha_m}^{p-1}$ to m^+ yields a nonzero vector m^- (of weight δ), mapped by f onto a multiple of v^-. In turn, applying $X_{\alpha_1}^{p-1} \ldots X_{\alpha_m}^{p-1}$ to m^- yields a nonzero vector m^{++} (of weight $(p-1)\delta$) mapped by f onto a multiple of v^+. Since m^{++} is killed by each X_α (cf. the statement dual to Theorem 5.1(a)), it is actually a maximal vector. It therefore generates a submodule of M which is a homomorphic image of $Z_{(p-1)\delta} = St$; but this submodule is mapped by f onto St, so we are done.

§6. Results of Verma and Hulsurkar

At this point we digress in order to summarize the work of Verma [2] and his student Hulsurkar [1] on the decomposition numbers d_λ defined in (5.2). This work contributes to the specific computations summarized in (5.2), but is not strictly necessary for the later theoretical results of the present paper. A word of caution: In the articles just cited, maps are usually written on the right rather than the left.

6.1 Verma's conjectures

Verma begins with Weyl's dimension polynomial $D(x_1, \ldots, x_\ell)$, which yields $dim(\lambda)$ for $\lambda = \Sigma(x_i - 1)\lambda_i$. For example, in type A_2,

$D(x_1,x_2) = x_1x_2(x_1+x_2)/2$. If $\sigma \in W$, let w_σ be the affine transformation of X described by $\lambda \mapsto \sigma\lambda + \delta_\sigma$, where $\delta_\sigma = \Sigma\lambda_i$ (summed over those i for which $\ell(\sigma_i\sigma) < \ell(\sigma)$). In turn, define a new polynomial $D_\sigma(x_1,\ldots,x_\ell) = D(w_\sigma(x))$. The following table lists these polynomials explicitly for type A_2, with the notation (r,s) in place of (x_1,x_2).

σ	δ_σ	$w_\sigma(r\lambda_1 + s\lambda_2)$	$2D_\sigma(r,s)$
1	(0,0)	(r,s)	rs(r+s)
σ_1	(1,0)	(1-r,r+s)	(1-r)(r+s)(1+s)
σ_2	(0,1)	(r+s,1-s)	(r+s)(1-s)(r+1)
$\sigma_1\sigma_2$	(1,0)	(1-r-s,r)	(1-r-s)r(1-s)
$\sigma_2\sigma_1$	(0,1)	(s,1-r-s)	s(1-r-s)(1-r)
σ_0	(1,1)	(1-s,1-r)	(1-s)(1-r)(2-r-s)

A quick calculation shows in this case that $\Sigma_\sigma D_\sigma(r,s) = 1$. More generally, Verma's <u>Conjecture</u> I reads:

There exist unique integers $b_\sigma (\sigma \in W)$ such that $\sum_{\sigma \in W} b_\sigma D_\sigma(\lambda) = 1$ for all $\lambda \in X$ (indeed, for all $\lambda \in E$).

(In the original formulation, the integers b_σ were required to be nonnegative, but examples in rank 3 showed that this was too restrictive.)

The polynomials D_σ are "harmonic" and take integral values for $\lambda \in X$. The space of W-harmonic polynomials had been shown by Steinberg to have dimension $|W|$, so the conjecture stated in particular that the D_σ must form a basis for this space. The proof of the conjecture by Hulsurkar [1] shows more precisely that the D_σ form a $\mathbf{Z}$-basis for the subgroup of W-harmonic polynomials taking integral values on X.

To indicate how this conjecture (now a theorem) is related to §5, we go further. Denote by $\lambda(\sigma)$ the weight in X_p linked to λ by σ, where λ lies in the lowest alcove of X_p. Thus $\lambda(\sigma) = \sigma(\lambda+\delta) + p\delta_\sigma - \delta$.

Verma's <u>Conjecture</u> <u>IV</u> states:

There exist integers b_σ ($\sigma \in W$), independent of p, such that $Z_\lambda \underset{\underline{u}}{\overset{\leftrightarrow}{=}} \underset{\sigma \in W}{\Sigma} b_\sigma \bar{V}_{\lambda(\sigma)}$ for λ belonging to the lowest alcove of X_p.

Once Conjecture I is proved, the integers b_σ found there can be inserted here and will at least insure that the indicated sum has the correct dimension $p^m = \dim Z_\lambda$. (Translation is now by $p\delta_\sigma$ rather than by δ_σ.) On the other hand, Verma observed in [2] that the truth of Conjecture IV for infinitely many values of p would imply the truth of Conjecture I. Subsequently he has informed the author that he has obtained a proof of Conjecture IV. (More recently, Jantzen has obtained a proof, along probably similar lines, by combining Theorem 9.2 below with results in Jantzen [4].)

Assume for the moment that integers b_σ are known which satisfy Conjecture IV (for all p). If the integers $a_{\lambda\mu}$ of (2.3) are also known, then in principle we can write down explicitly the decompositions: $Z_\lambda \underset{\underline{u}}{\overset{\leftrightarrow}{=}} \Sigma d_\mu M_\mu$. Moreover, if the $a_{\lambda\mu}$ are known to be essentially independent of p (i.e., to be dependent just on the position of λ, μ in their linkage class), then the same must be true of the d_μ. These remarks lead directly to the results listed in (5.2) for small ranks.

6.2 Some special cases

Verma verified his Conjecture I directly in some cases of low rank, obtaining specific values for the integers b_σ. (Later Hulsurkar computed the b_σ in other cases; some of these turned out to be negative.) For reference we list the numbers given in Verma [2]. As in (5.2), the alcoves are indicated schematically, with the number b_σ placed alongside the alcove linked to the <u>top</u> one by σ:

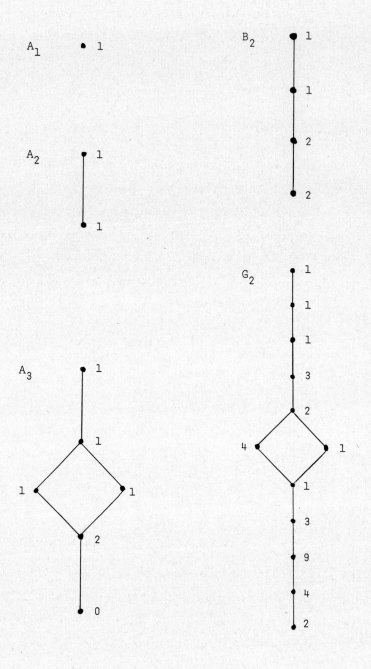

It is noteworthy that these numbers b_σ are actually invariants attached to the alcoves, rather than to the elements of W (e.g., for type A_2 there are only two of them to find, not six). Moreover, the number attached to the top alcove is always 1, as follows from the proof of Hulsurkar to be sketched below. The numbers b_σ seem otherwise to behave in no predictable way.

6.3 Hulsurkar's method of proof

It may be useful to sketch in a couple of special cases the proof of Verma's Conjecture I given by Hulsurkar. His idea is "to try to solve a set of $|W|$ equations obtained from Verma's conjectural formula $\Sigma b_\tau D_\tau(x) = 1$ by substituting a judiciously chosen set of $|W|$ elements of X in place of x". The weights chosen are of the form $-\varepsilon_{\sigma_0 \sigma} = -\sigma^{-1}\sigma_0(\delta_{\sigma_0 \sigma})$, where $\sigma \in W$. For type A_2 these weights are as follows:

σ	$-\varepsilon_{\sigma_0 \sigma}$
1	(1,1)
σ_1	(0,1)
σ_2	(1,0)
$\sigma_1 \sigma_2$	(1,-1)
$\sigma_2 \sigma_1$	(-1,1)
σ_0	(0,0)

In this case the resulting matrix $(D_\sigma(-\varepsilon_{\sigma_0 \tau}))_{\sigma,\tau \in W}$ is simply the identity matrix; it is in particular unipotent, and the column sums of its inverse give the desired numbers b_σ, here all equal to 1. In general, Hulsurkar shows that the matrix in question is unipotent (which leads him to suggest a new partial ordering of W). The proof shows that column sums in the inverse matrix which correspond to the top alcove must yield $b_\sigma = 1$, as mentioned earlier. We shall give the computation for type B_2 (where α_1 is long, α_2 short). Recall that $D(r,s) = rs(r+s)(2r+s)/6$.

σ	δ_σ	$-\varepsilon_{\sigma_0}\sigma$	$6D_\sigma(r,s)$	b_σ
1	$(0,0)$	$(1,1)$	$rs(r+s)(2r+s)$	2
σ_1	$(1,0)$	$(0,1)$	$(1-r)(2r+s)(1+r+s)(2+s)$	1
σ_2	$(0,1)$	$(1,0)$	$(r+s)(1-s)(1+r)(1+2r+s)$	1
$\sigma_1\sigma_2$	$(1,0)$	$(1,-1)$	$(1-r-s)(2r+s)(1+r)(2-s)$	1
$\sigma_2\sigma_1$	$(0,1)$	$(-1,2)$	$(r+s)(1-2r-s)(1-r)(1+s)$	2
$\sigma_1\sigma_2\sigma_1$	$(1,0)$	$(-1,1)$	$(1-r-s)s(1-r)(2-2r-s)$	2
$\sigma_2\sigma_1\sigma_2$	$(0,1)$	$(1,-2)$	$r(1-2r-s)(1-r-s)(1-s)$	2
σ_0	$(1,1)$	$(0,0)$	$(1-r)(1-s)(2-r-s)(3-2r-s)$	1

$$(D_\sigma(-\varepsilon_{\sigma_0}\tau)) = \begin{pmatrix} 1 & 0 & 0 & 0 & 0 & 0 & 0 & 0 \\ 0 & 1 & 0 & 0 & 0 & -1 & 0 & 0 \\ 0 & 0 & 1 & 0 & 0 & 0 & -1 & 0 \\ -1 & 0 & 0 & 1 & 0 & 0 & 0 & 0 \\ & & & & 1 & 0 & 0 & 0 \\ & & O & & 0 & 1 & 0 & 0 \\ & & & & 0 & 0 & 1 & 0 \\ & & & & -1 & 0 & 0 & 1 \end{pmatrix}$$

inverse matrix =
(column sum
yields b_σ)

$$\begin{pmatrix} 1 & & & & 0 & & & \\ & 1 & & & & 1 & & \\ & & 1 & & & & 1 & \\ 1 & & & 1 & & & & 0 \\ & & & & 1 & & & \\ & & O & & & 1 & & \\ & & & & & & 1 & \\ & & & & 1 & & & 1 \end{pmatrix}$$

§7. Tensor products

Tensor products will play a key role in later sections. Here we assemble a few basic facts.

7.1 Weights in a tensor product

PROPOSITION. Let M, N be G-modules generated by maximal vectors v, w of respective weights $\lambda, \mu \in X^{+}$. Then $v \otimes w$ is (up to scalar multiples) the unique maximal vector of weight $\lambda + \mu$ in $M \otimes N$, and all weights ν of $M \otimes N$ satisfy $\nu \leq \lambda + \mu$. Moreover, $\mathrm{ch}(M \otimes N) = \mathrm{ch}(M) \cdot \mathrm{ch}(N)$.

All of this is well known and easy to verify. The proposition applies in particular to the modules $\bar{V}_\lambda$, M_λ ($\lambda \in X^{+}$).

7.2 Composition factors

In characteristic 0 the tensor product $V_\lambda \otimes V_\mu$ is isomorphic to a direct sum of various V_ν, $\nu \leq \lambda + \mu$, with V_ν occurring (say) $c(\lambda,\mu,\nu)$ times. Formally, $\mathrm{ch}(\lambda) \cdot \mathrm{ch}(\mu) = \sum_\nu c(\lambda,\mu,\nu) \mathrm{ch}(\nu)$, with $c(\lambda,\mu,\lambda+\mu) = 1$. There are explicit algorithms for computing these multiplicities (e.g., the formula of Steinberg, cf. Humphreys [3, 24.4]).

In characteristic p we can take minimal admissible lattices in V_λ and V_μ, then tensor them to get an admissible lattice in $V_\lambda \otimes V_\mu$. Comparing formal characters, we see that $\bar{V}_\lambda \otimes \bar{V}_\mu \underset{G}{\leftrightarrow} \overline{V_\lambda \otimes V_\mu} \underset{G}{\leftrightarrow} \sum c(\lambda,\mu,\nu) \bar{V}_\nu$. In case we know all the numbers $a_{\lambda\mu}$ of (2.3), we can obtain recursively the composition factors of $M_\lambda \otimes M_\mu$. It would also be possible in principle to express $p\text{-}\mathrm{ch}(\lambda) \cdot p\text{-}\mathrm{ch}(\mu)$ directly as a sum of certain $p\text{-}\mathrm{ch}(\nu)$, but in practice this is very difficult except in rank 1.

Problems:

(1) Find a closed formula, comparable to Steinberg's, for the multiplicity of a composition factor of $M_\lambda \otimes M_\mu$.

(2) Determine the indecomposable summands of $M_\lambda \otimes M_\mu$. (Linkage classes of weights yield only a partial decomposition.)

There is another approach which sometimes yields very precise information about tensor products with a minimum of computation. It is known in characteristic 0 (cf. Brauer [1], Humphreys [3, Exercises 24.9, 24.12], Jantzen [2, p. 131]) that $ch(\lambda) \cdot ch(\mu) = \sum_\nu m_\lambda(\nu) ch(\nu+\mu)$, where $ch(\pi)$ is defined formally to be 0 if $\pi + \delta$ is irregular or to be $(\det \sigma) ch(\sigma(\pi+\delta)-\delta)$ if $\pi + \delta$ is regular and $\sigma(\pi + \delta)$ is dominant. In particular, if all $\nu + \mu$ occurring are dominant, it is easy to read off the constituents of $V_\lambda \otimes V_\mu$ once the weight multiplicities of V_λ are known. In any case, the only possible highest weights of irreducible constituents of $V_\lambda \otimes V_\mu$ are of the form $\nu + \mu$ (ν a weight of V_λ). The formal calculation here carries over to the case in which $ch(\lambda)$ is replaced by an arbitrary element of $\mathbb{Z}[X]^W$, e.g., by $p\text{-}ch(\lambda)$ (cf. Jantzen, loc. cit.).

PROPOSITION. Let $\lambda, \mu \in X^+$. If $p\text{-}ch(\lambda) = \sum_\nu m'_\lambda(\nu) e(\nu)$, then $p\text{-}ch(\lambda) \cdot ch(\mu) = \sum_\nu m'_\lambda(\nu) ch(\nu+\mu)$.

In particular, the G-composition factors of $M_\lambda \otimes \bar{V}_\mu$ can be written down explicitly once we know those of the $\bar{V}_{\nu+\mu}$.

7.3 Twisted tensor products

Besides ordinary tensor products, we have to consider twisted tensor products. Recall that each $\lambda \in X^+$ has a unique expression $\lambda = \mu_0 + p\mu_1 + \ldots + p^k \mu_k$ ($\mu_i \in X_p$); in turn, Steinberg proved that $M_\lambda \cong M_{\mu_0} \otimes M_{\mu_1}^{(p)} \otimes \ldots \otimes M_{\mu_k}^{(p^k)}$ (2.1).

PROPOSITION. Let $\lambda \in X^{+}$ be as above.

(a) Viewed as a $K\Gamma$-module, M_{λ} is isomorphic to
$M_{\mu_0} \otimes M_{\mu_1} \otimes \cdots \otimes M_{\mu_k}$.

(b) Viewed as a $\underline{u}$-module, M_{λ} is isomorphic to a direct sum of copies of M_{μ_0}, p-dim$(\mu_1) \cdots$ p-dim(μ_k) in number.

Proof. Part (a) is clear, while part (b) is well known (cf. Humphreys [1, p. 73] or Ballard [1, pp. 87-88].

Combined with (7.2), the proposition affords a recursive method for computing the $K\Gamma$-composition factors of any G-module $M_{\lambda} \otimes M_{\mu}$ ($\lambda, \mu \in X^{+}$), once all $a_{\lambda\mu}$ are known.

§8. Construction of projective modules

8.1 Tensoring with a projective module

The following lemma will enable us to construct new projective modules from known ones.

LEMMA. Let A be the group algebra (over a field) of a finite group, or the restricted universal enveloping algebra of a Lie p-algebra. If M and P are (left) A-modules, with P projective, then M⊗P is a projective (left) A-module.

Proof. For the case of a group algebra, cf. Curtis and Reiner [1, p. 426, ex. 2]. For the case of a Lie p-algebra, cf. Pareigis [1, Lemma 2.5] or Humphreys [4]. A more comprehensive result is sketched in Appendix T.

8.2 The main theorem

The only PIM for $\underline{u}$ which we know explicitly at this point is the Steinberg module St = $M_{(p-1)\delta}$, cf. (5.5). For each $\mu\epsilon\Lambda$, form the new $\underline{u}$-module T_μ = $M_\mu \otimes$ St. Thanks to Lemma 8.1, T_μ is projective. We can also regard T_μ as a G-module (hence as a KΓ-module, cf. §10). Its weights relative to T are certain $\nu \leq \mu + (p-1)\delta$, $\mu + (p-1)\delta$ itself occurring with multiplicity 1 (7.1). As a projective $\underline{u}$-module, T_μ can be written as a direct sum of certain PIM's Q_λ, each having a well determined multiplicity. For reasons of dimension, we can expect in general to have many such summands; but only one of them can intersect the highest weight space. We shall prove that this Q_λ occurs just once in T_μ, and is a G-module direct summand.

The following theorem was announced in Humphreys, Verma [1]. It was inspired by work of Jeyakumar [1] for SL(2,q), as reformulated in Humphreys [4].

THEOREM. Fix $\mu \epsilon \Lambda$ and set $\lambda = (\mu-\delta)^\circ$. Then Q_λ occurs precisely once as a $\underline{u}$-summand of T_μ = $M_\mu \otimes$ St, and is also a G-summand.

Proof. Choose a weight vector v^+ in T_μ of weight $\mu + (p-1)\delta$, relative to T; v^+ is unique up to a scalar multiple. Its weight relative to $\underline{t}$ is just $\mu - \delta = \lambda^\circ$. Let Z be the $\underline{u}$-submodule of T_μ generated by v^+. Since v^+ is a maximal vector for G, hence for $\underline{u}$, Z is a homomorphic image of Z_{λ° (5.1). In view of (1.2), Z is spanned by weight vectors relative to T. Their weights $\nu \epsilon$ X all satisfy:

(*) $\qquad \nu \geq \mu + (p-1)\delta - 2(p-1)\delta$,

since the unique (up to scalars) minimal vector of Z_{λ° is gotten by applying $Y_{\alpha_1}^{p-1} \ldots Y_{\alpha_m}^{p-1}$ to a generator, cf. Theorem 5.1(a). Moreover, equality holds in (*) for some ν if and only if Z is isomorphic to Z_{λ^o}, cf. Theorem 5.1(c). We shall show that this does happen.

Since each Q_ν occurring as a summand of T_μ has a unique irreducible submodule (isomorphic to M_ν), it is clear that every irreducible sub-module of T_μ lies in the sum of these. This applies in particular to any irreducible submodule of Z. But the second corollary in (5.4) implies that Q_ν has a submodule isomorphic to Z_{ν^o}, so a minimal vector of its irreducible submodule M_ν must be obtainable by applying the monomial $Y_{\alpha_1}^{p-1} \ldots Y_{\alpha_m}^{p-1}$ to some other vector. From this it follows that a minimal vector of an irreducible submodule of Z must be a sum of weight vectors relative to T of weights $\nu \in X$ satisfying:

(**) $\nu \leq \mu + (p-1)\delta - 2(p-1)\delta.$

Taken together, (*) and (**) imply that Z is isomorphic to Z_{λ^o}, as claimed. Moreover, the unique irreducible $\underline{u}$-submodule (of type M_λ) in Z is actually generated by a weight vector relative to T of <u>restricted</u> weight $\lambda \in X_p$.

In view of (5.4), the preceding argument shows that Q_λ does occur at least once as a $\underline{u}$-summand of T_μ. We claim that it cannot occur more than once. This can be seen concretely by examining weights relative to T, but there is also an easy abstract argument, as follows.

The multiplicity with which Q_λ occurs in a direct sum decomposition of T_μ can be measured (cf. (R.2)) by the dimension of $\mathrm{Hom}_{\underline{u}}(M_\mu \otimes St, M_\lambda)$. But this space is canonically isomorphic to the space of $\underline{u}$-invariants in $M_\lambda \otimes (M_\mu \otimes St)^* \cong M_\lambda \otimes (M_\mu^* \otimes St) \cong (M_\lambda \otimes M_\mu^*) \otimes St \cong \mathrm{Hom}(St, M_\lambda \otimes M_\mu^*)$. In turn, the dimension of $\mathrm{Hom}_{\underline{u}}(St, M_\lambda \otimes M_\mu^*)$ measures the multiplicity of St as a composition factor of $M_\lambda \otimes M_\mu^*$. Since $M_\mu^* \cong M_{-\sigma_o \mu}$ (2.4), while $\lambda = (\mu-\delta)^o$, the highest weight of this tensor product is $\lambda + (-\sigma_o\mu) = (p-1)\delta$. So St can occur only once here.

The argument shows that Q_λ occurs just once as a $\underline{u}$-summand in any decomposition of T_μ, hence that T_μ has a unique $\underline{u}$-submodule of type M_λ. It follows that the sum of all $\underline{u}$-submodules isomorphic to Q_λ is indecomposable. But the injective property of any such Q_λ allows us to split it off as a direct summand. The conclusion is that Q_λ occurs just once as a $\underline{u}$-submodule of T_μ. Moreover, it has a unique complement in T_μ, the submodule gotten by adding up all PIM's in T_μ other than Q_λ.

It remains to consider how G acts on T_μ. Since G acts on $\underline{g}$ as a group of automorphisms (via the adjoint representation), it is clear that $g \cdot Q_\lambda$ ($g \in G$) is another PIM for $\underline{u}$. To see which PIM it is, let g be a standard unipotent generator of G corresponding to a positive (resp. negative) root, and observe that g sends a maximal (resp. minimal) vector in the submodule $M_\lambda \subset Q_\lambda$ to another such vector of the same weight, forcing $g \cdot Q_\lambda = Q_\lambda$ in view of the preceding paragraph. The same conclusion then holds for arbitrary $g \in G$. (This type of argument is due to Curtis [1, pp. 317-318], cf. Humphreys [1, p. 73].) Similarly, G stabilizes the sum of all PIM's in T_μ not of type Q_λ, so Q_λ is a G-summand as required.

8.3 Complements

There are a few remarks to be made about Theorem 8.2 and its proof.

(a) The proof does not require the linkage principle.

(b) We did not show that each PIM occurring in T_μ must be G-stable, only that the sum of all those of a given type Q_λ must be so.

(c) Verma has raised the question whether a given indecomposable $\underline{u}$-module (such as Q_λ) can have distinct G-module structures yielding the same derived action of $\underline{u}$. He suggests that the answer may well be no. (Indecomposability is of course crucial here.)

(d) In a related vein, one can ask for an intrinsic characterization of the G-modules Q_λ which would single them out from the collection of all indecomposable G-modules.

(e) It seems to be possible to avoid the use of the "Corollary of proof" (5.4), by relying instead on a straightforward analogue for $\underline{u}$ of Dixmier [1, 7.6.14]. This would make the proof of Theorem 8.2 somewhat more self-contained.

8.4 Mumford's conjecture

The technique of tensoring with the Steinberg module may provide an interesting variation on the proof of Mumford's conjecture given by W. J. Haboush [1]. The idea is as follows. We are given a G-module M containing a 1-dimensional G-submodule L (necessarily of highest weight 0), and we are to prove the existence of a G-module homomorphism M → End(St_n), nonzero on L, for sufficiently large n. (This is the key step in Haboush's proof that G is "geometrically reductive".)

An equivalent problem is to prove the existence of a G-homomorphism $M \otimes St_n$ → St_n, nonzero on $L \otimes St_n$ ($\tilde{=} St_n$). This amounts to showing that $L \otimes St_n$ is a _direct_ _summand_ of the tensor product. To investigate the indecomposable G-summands of $M \otimes St_n$, we compare the action of G with that of $\underline{u}_n$ (cf. Appendix U). Since St_n is a projective $\underline{u}_n$-module, $M \otimes St_n$ breaks up into a direct sum of PIM's $Q_{\lambda,n}$, and the sum of those occurring for a fixed λ is a G-summand. Choose n so large that all weights of M are below $(p^n-1)\delta$. Then St_n is the only G-composition factor of $M \otimes St_n$ which becomes a sum of copies of St_n on restriction to $\underline{u}_n$. So the G-summand of the tensor product corresponding to the PIM St_n has G-composition factors all isomorphic to St_n, and $(p^n-1)\delta$ is the only highest weight occurring. It follows that G acts completely reducibly here, allowing us to split off the G-submodule $L \otimes St_n$ as a direct summand.

This approach seems to be fairly straightforward and noncomputa-
tional, once one has established the results listed in Appendix U.

§9. Small PIM's

In this section (L) is essential.

The G-module $M_\mu \otimes St$ considered in (8.2) can always be decomposed
into a direct sum of G-submodules corresponding to the various linkage
classes involved. When μ is "small", i.e., located in the lowest
alcove, the G-summand belonging to the linkage class of $\mu + (p-1)\delta$ will
turn out to be just Q_λ, $\lambda = (\mu - \delta)^\circ$ in Λ (with perhaps a mild restriction on p). In this case it will be easy to compute the formal
character of Q_λ.

9.1 Orbits of weights

If $\mu \in X$, we denote by W_μ its stabilizer in W. Denote by W^μ some
fixed (but arbitrary) set of coset representatives for W/W_μ.

LEMMA. Let $\mu \in X_p$ lie in the lowest alcove, i.e., $<\mu,\alpha> < p$ for
all $\alpha \in \Phi^+$. Assume that p does not divide f. Then:

(a) If ν is a weight of M_μ and $\nu \equiv \mu \pmod{pX}$, then $\nu = \mu$.

(b) The W-orbit of μ in Λ has the same cardinality $[W:W_\mu]$ as the
W-orbit of μ in X.

Proof. (a) This is proved in Ballard [1, Lemma 2 of §8, cf. 4.3],
where the stated assumption that p exceeds f is not needed, only the
weaker assumption we have made. (Alternatively, one can argue in a
slightly different way by using remarks of Verma [2, §5] concerning
the relationship between his affine Weyl group and the larger group
obtained by allowing arbitrary translations by elements of pX.)

(b) It is well known that W_μ is generated by the reflections σ_α
which it contains; when p does not divide f, the same is true of the

stabilizer of μ regarded as an element of Λ (cf. Verma [2, §6]; this has also been observed by Steinberg). Now if σ_α fixes μ in Λ, it follows that $\langle\mu,\alpha\rangle \equiv 0 \pmod{p}$. By hypothesis, the absolute value of $\langle\mu,\alpha\rangle$ is less than p, so $\langle\mu,\alpha\rangle = 0$ and $\sigma_\alpha \in W_\mu$.

In the context of part (b) of the lemma, let $\lambda = (\mu - \delta)^\circ$ in Λ. Then the cardinality of the W-orbit of μ in Λ clearly equals the cardinality of the linkage class of λ in Λ, which we have denoted a_λ.

9.2 Main theorem

Now we can state the main result of this section.

THEOREM. Let $\mu \in \Lambda$ lie in the lowest alcove, let $\lambda = (\mu-\delta)^\circ$ in Λ, and construct Q_λ in $M_\mu \otimes \mathrm{St}$ as in (8.2). Assume that p does not divide f. Then:

(a) Q_λ involves all the composition factors of $M_\mu \otimes \mathrm{St}$ having highest weight linked to λ.

(b) $Q_\lambda \overset{\leftrightarrow}{G} \underset{\sigma \in W^\mu}{\Sigma} \bar{V}_{\sigma\mu+(p-1)\delta}$.

(c) The dimension of Q_λ is $[W:W_\mu]p^m = a_\lambda p^m$, so in particular $d_\lambda = 1$.

(d) The formal character of Q_λ is $s(\mu) \cdot \mathrm{ch}((p-1)\delta)$, where $s(\mu)$ is the sum of the $e(\sigma\mu)$, $\sigma \in W^\mu$.

Proof. Since μ is small, $\bar{V}_\mu = M_\mu$ (4.1). Moreover, $|\langle\sigma\mu,\alpha\rangle| < p$ for all roots α, so that all $\sigma\mu + (p-1)\delta$ are dominant. Thanks to (7.2), $M_\mu \otimes \mathrm{St} \overset{\leftrightarrow}{G} \Sigma \bar{V}_\pi$, $\pi \in X^+$ running over certain $\nu + (p-1)\delta$ for which ν is a weight of M_μ, including all $\sigma\mu + (p-1)\delta$ (each occurring once). Apart from the weights $\sigma\mu + (p-1)\delta$, we assert that no such weight can be linked to λ in X. Say $\sigma(\lambda+\delta) = \nu + (p-1)\delta + \delta = \nu + p\delta$ in Λ. Since $\sigma_o\mu = \lambda + \delta$ in Λ, this implies that ν is W-conjugate to $\mu\pmod{pX}$. Then Lemma 9.1(a) shows that ν is W-conjugate to μ.

As a result, all those composition factors of the G-module $M_\mu \otimes St$ having highest weight linked to λ occur in the $\bar{V}_\pi$, $\pi = \sigma\mu + (p-1)\delta$ ($\sigma \in W^\mu$), the latter each occurring once in the tensor product. Let us use Weyl's character formula (see, for example, Humphreys [3, 24.3]) to compute the formal character of the sum of these $\bar{V}_\pi$, using q as an abbreviation for the denominator $\sum_{\tau \in W} \epsilon(\tau)e(\tau\delta)$, ϵ being the alternating character of W:

$$\sum_{\sigma \in W^\mu} ch(\sigma\mu + (p-1)\delta) =$$

$$\sum_{\sigma \in W^\mu} \sum_{\tau \in W} \epsilon(\tau)e(\tau(\sigma\mu + p\delta))/q =$$

$$\sum_{\sigma \in W^\mu} \sum_{\tau \in W} \epsilon(\tau)e(\tau\sigma\mu)e(p\tau\delta)/q =$$

$$(\sum_{\sigma \in W^\mu} e(\sigma\mu))(\sum_{\tau \in W} \epsilon(\tau)e(p\tau\delta)/q).$$

The first factor is what we have denoted $s(\mu)$, while the second factor is just the formal character $ch((p-1)\delta)$ of the Steinberg module.

The dimension of the λ-linked component of $M_\mu \otimes St$ is therefore $[W:W_\mu]p^m$, and of course it includes Q_λ. But we already know from (5.4) that $\dim Q_\lambda = a_\lambda d_\lambda p^m$. The concluding remark in (9.1), along with Lemma 9.1(b), shows that $a_\lambda = [W:W_\mu]$, and from this all assertions of the theorem follow immediately.

(The formal character computation above was found independently by Ballard [1, §5, §8].)

What can be said when p does divide f? Although the picture is not entirely clear, it seems likely that most of the theorem will remain valid, provided we compensate for the possibility that the W-orbit of μ in Λ (or the linkage class of λ in Λ) may be smaller than expected. Then the number d_λ should be p, and $s(\mu)$ should involve repetitions. See Verma [2, §7] for some discussion of this "ramification" phenomenon. In place of (9.1), Lemma 3 of Jantzen [3] ought to be relevant here.

9.3 An example: SL(3,K)

In this subsection $G = SL(3,K)$. We shall describe the composition factors of the G-module $Q_\lambda \subset M_\mu \otimes St$ considered in (9.2), where μ belongs to the lowest alcove. Assume that $p \neq 3$, to avoid ramification.

We have seen that $Q_\lambda \underset{G}{\leftrightarrow} \underset{\sigma \in W^\mu}{\Sigma} \bar{V}_{\sigma\mu+(p-1)\delta}$. A knowledge of how to express the various ch(ν) in terms of various p-ch(ν) ($\nu \in X^+$) would in principle suffice to determine the G-composition factors. (When $G = SL(2,K)$, there are so few linked weights to consider that the answer can be written down effortlessly, cf. Humphreys [4]. Here a little more work is needed.) Consider the dominant weights $\leq \mu+(p-1)\delta$ and linked to this weight. Since μ lies in the lowest alcove, a glance at the weight region shows that $\mu + (p-1)\delta$ lies in the alcove

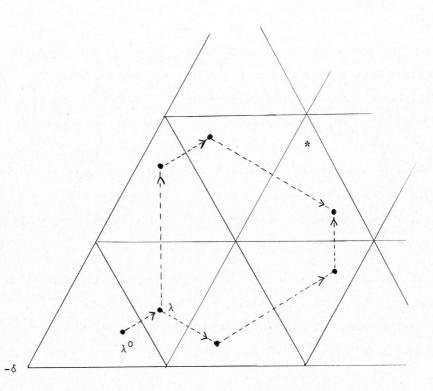

labelled * and also that there are just six lower linked weights, as
indicated. We can exhibit in a table the precise weights involved,
assuming for simplicity that λ lies in the interior of the top alcove
(so that dim $Q_\lambda = 6p^3$). The degenerate cases can be handled similarly.
Set $\lambda = (r,s)$, $\mu = (p-s-1, p-r-1)$, $\lambda(\sigma) =$ weight in Λ linked to λ by σ.
Observe that the fourth and sixth columns of the table contain the
same weights, but in reversed order.

For each weight ν in the last column of the table, we know that
$\bar{V}_\nu$ and hence M_ν does figure in Q_λ. The seventh linked weight in the
diagram is just λ°, and this too figures in Q_λ since $\bar{V}_\lambda \underset{G}{\leftrightarrow} M_\lambda \oplus M_{\lambda^\circ}$.
All that remains to be settled is the <u>multiplicity</u> of each
composition factor of Q_λ. For this, we can view Q_λ as a <u>u</u>-module.
Recall (5.4) that the <u>u</u>-composition factors of Q_λ are the $M_{\lambda(\sigma)}$, with
multiplicity $6d_{\lambda(\sigma)}$ (= 6 or 12 depending on whether the linked weight
is in the upper or lower alcove). In view of Proposition 7.3(b), and
the fourth column of the above table, we obtain the G-composition
factors of Q_λ by the following recipe:

σ	tensor $M_{\lambda(\sigma)}$ with:
1	$6M_{00}^{(p)}$
σ_1	$4M_{10}^{(p)}$
σ_2	$4M_{01}^{(p)}$
$\sigma_1\sigma_2$	$2M_{10}^{(p)}$
$\sigma_2\sigma_1$	$2M_{01}^{(p)}$
$\sigma_\circ$	$4M_{00}^{(p)} \oplus M_{11}^{(p)}$

Observe that the dimension of the module tensored with $M_{\lambda(\sigma)}$ is
just $6d_{\lambda(\sigma)}$. This suggests that, if we had not already known the
values of the d_λ, we could have discovered them by working out the
p-ch(ν) in terms of the ch(ν).

σ	δ_σ	$\lambda(\sigma)$	$\lambda(\sigma)+p\delta_\sigma$	$\sigma\mu$ (in X)	$\sigma\mu + (p-1)\delta$
1	(0,0)	(r,s)	(r,s)	(p-s-1,p-r-1)	(2p-s-2,2p-r-2)
σ_1	(1,0)	(p-r-2,r+s+1-p)	(2p-r-2,r+s+1-p)	(-p+s+1,2p-r-s-2)	(s,3p-r-s-3)
σ_2	(0,1)	(r+s+1-p,p-s-2)	(r+s+1-p,2p-s-2)	(2p-r-s-2,-p+r+1)	(3p-r-s-3,r)
$\sigma_1\sigma_2$	(1,0)	(2p-r-s-3,r)	(3p-r-s-3,r)	(-2p+r+s+2,p-s-1)	(-p+r+s+1,2p-s-2)
$\sigma_2\sigma_1$	(0,1)	(s,2p-r-s-3)	(s,3p-r-s-3)	(p-r-1,-2p+r+s+2)	(2p-r-2,-p+r+s+1)
σ_0	(1,1)	(p-s-2,p-r-2)	(2p-s-2,2p-r-2)	(-p+r+1,-p+s+1)	(r,s)

§10. PIM's of KΓ

10.1 The Steinberg module

For each $\lambda \in X_p$, R_λ denotes the PIM of KΓ having M_λ as its unique top (and bottom) composition factor. Recall (R.1) that each projective KΓ-module is a direct sum of PIM's. Moreover, dim R_λ is a multiple of p^m (the exact power of p dividing $|\Gamma|$). Since dim $M_\lambda < p^m$ unless $\lambda = (p-1)\delta$, cf. (2.2), the Steinberg module St = $M_{(p-1)\delta}$ is the only irreducible KΓ-module which could possibly be projective. In fact:

PROPOSITION. St is a projective KΓ-module.

Proof. The only proof of this statement known to the author requires a highly nontrivial fact about the ordinary representations of Γ: There exists an irreducible representation of Γ over $\mathbb{C}$ (indeed over Q) of degree p^m. This was discovered by Steinberg [2; 4, 15.5], cf. Appendix S. In turn, Brauer's theory of blocks of defect 0 (Curtis, Reiner [1, Theorem 86.3], Dornhoff [1, Part B, §62]) implies that this representation remains irreducible upon reduction modulo p and is then a PIM, which for reasons of dimension must be St.

10.2 Comparison of Q_λ and R_λ

Return now to the situation of Theorem 8.2, with Q_λ contained in $M_\mu \otimes St$ precisely once as a G-summand. In view of Lemma 8.1 and Proposition 10.1, the tensor product $M_\mu \otimes St$ is projective as a KΓ-module; hence the Γ-summand Q_λ is projective also. But M_λ is a G-submodule of Q_λ, whose restriction to Γ is isomorphic to the irreducible KΓ-module M_λ. Since a PIM R_ν has M_ν as its unique irreducible submodule, we conclude that R_λ must occur as a Γ-summand of Q_λ. This proves:

THEOREM. Let $\lambda \in X_p$. The G-module Q_λ of (8.2) is a projective $K\Gamma$-module, having R_λ as a direct summand. In particular, $\dim R_\lambda \le \dim Q_\lambda$.

This theorem was announced in Humphreys, Verma [1]; it was suggested by the case $\Gamma = SL(2,p)$ which had been treated by Jeyakumar [1] (cf. Humphreys [4]). Essentially the same result has been obtained independently by Ballard [1], who emphasizes the Brauer characters of the PIM's of $K\Gamma$. He is able to show (Corollary 1 of Theorem 2, p. 41) that R_λ occurs just once in $M_\mu \otimes St$; therefore it occurs only once in Q_λ. It should be pointed out that our Q_λ is essentially the same thing as Ballard's $Q(f(\mu))$, although he does not consider the PIM's of $\underline{u}$. On the other hand, his methods give some information about the PIM's of $\Gamma_n = G(\mathbb{F}_{p^n})$. From our point of view, these are more naturally approached via twisted tensor products, by analogy with the irreducible modules.

COROLLARY. Let $\lambda \in X_q$, $q = p^n$, and denote by $R_{\lambda,n}$ the PIM of $K\Gamma_n$ corresponding to M_λ. Write $\lambda = \nu_0 + p\nu_1 + \ldots + p^{n-1}\nu_{n-1}$, $\nu_i \in X_p$. Set $Q_{\lambda,n} = Q_{\nu_0} \otimes Q_{\nu_1}^{(p)} \otimes \ldots \otimes Q_{\nu_{n-1}}^{(p^{n-1})}$, viewed as a G-module. Then $Q_{\lambda,n}$ is a projective $K\Gamma_n$-module, having $R_{\lambda,n}$ as a direct summand.

Proof. As in (10.1), the Steinberg module $M_{(q-1)\delta}$ for $K\Gamma_n$ is projective, call it St_n. Define $\mu_i = (p-1)\delta + \sigma_0\nu_i$, $\mu = \mu_0 + p\mu_1 + \ldots + p^{n-1}\mu_{n-1}$. Then Lemma 8.1 shows that $M_\mu \otimes St_n$ is a projective $K\Gamma_n$-module. On the other hand, $M_\mu \otimes St_n \cong (M_{\mu_0} \otimes St) \otimes (M_{\mu_1} \otimes St)^{(p)} \otimes \ldots \otimes (M_{\mu_{n-1}} \otimes St)^{(p^{n-1})}$ as G-modules. The right side contains $Q_{\lambda,n}$ as a G-summand, in view of (8.2), with M_λ occurring as a G-submodule. Therefore $Q_{\lambda,n}$ is a projective $K\Gamma_n$-module, as asserted, involving $R_{\lambda,n}$ as a direct summand. (The notation $R_{\lambda,n}$ or $Q_{\lambda,n}$ is used because R_λ or Q_λ alone would be ambiguous.)

For SL(2,q), cf. Jeyakumar [1], Humphreys [4]. The corollary was
announced in Humphreys, Verma [1].

The parallel between $K\Gamma$ and $\underline{u}$ can be extended to one between $K\Gamma_n$
and an associative algebra $\underline{u}_n$, as was pointed out several years ago
by Verma in a letter to the author. Begin with the Kostant Z-form U_Z
of the universal enveloping algebra of g_C, and set $U_K = U_Z \otimes K$. Then
define $\underline{u}_n$ to be the subalgebra of U_K generated by all
$(X_\alpha^k/k!) \otimes 1$, $(Y_\alpha^k/k!) \otimes 1$, where $k < p^n = q$. It can be seen that $\underline{u}_1 = \underline{u}$
and that $\dim \underline{u}_n = q^{\dim g}$. Further, Verma observed that the irredu-
cible modules (resp. PIM's) for $\underline{u}_n$ are just the M_λ, $\lambda \in X_q$ (resp.
$Q_{\lambda,n}$, $\lambda \in X_q$). (See Appendix U.)

10.3 Regularity conjecture

Since R_λ ($\lambda \in X_p$) occurs as a $K\Gamma$-summand of Q_λ, it is natural to
ask when R_λ equals Q_λ. Verma suggested the following answer:

CONJECTURE. Let $\lambda \in X_p$, $\lambda = \Sigma c_i \lambda_i$ ($0 \le c_i < p$). If λ is regular,
i.e., if all $c_i \ne 0$, then $R_\lambda = Q_\lambda$ (viewed as $K\Gamma$-modules).

One might also conjecture (as in Humphreys, Verma [1]) that
regularity is a necessary condition for equality to hold; but at least
for twisted groups there exist counterexamples, cf. (15.3) below.

We shall review here some of the evidence supporting the
conjecture. To put the conjecture into perspective, recall (R.2) that
$K\Gamma$ can be written as a direct sum of PIM's, with R_λ occurring $\dim M_\lambda$
times; for $\underline{u}$ there is an analogous decomposition. Therefore
$$|\Gamma| = \dim K\Gamma = \sum_{\lambda \in X_p} (\dim M_\lambda)(\dim R_\lambda) \le \sum_{\lambda \in \Lambda} (\dim M_\lambda)(\dim Q_\lambda) = \dim \underline{u}$$
$= p^{2m+\ell}$. But $|\Gamma|$ is a polynomial in p whose highest term is precisely
$p^{2m+\ell}$ (as is well known). Qualitatively speaking, it is therefore
plausible that we should have $R_\lambda = Q_\lambda$ for "most" λ. On the other hand,
there is good reason to believe that $R_\lambda \ne Q_\lambda$ occurs frequently when λ
is irregular.

(a) The weight $\lambda = (p-1)\delta$ is highly regular, and indeed we know already that $R_\lambda = St = Q_\lambda$ in this case.

(b) In case $\Gamma = SL(2,p)$, it was shown by Brauer, Nesbitt [1] that all R_λ have dimension 2p except those corresponding to $\lambda = 0$, $p-1$. On the other hand, dim $Q_\lambda = 2p$ unless $\lambda = p-1$ (5.4). Direct examination of the tensor product (as in Humphreys [4]) makes it clear that $Q_0 = R_0 \oplus R_{p-1}$. So the conjecture holds in this case, along with its converse.

(c) Besides $SL(2,p)$, there are a few other small cases for which the Cartan p-invariants of Γ have been calculated in an ad hoc manner (cf. Humphreys [5]): SL(3,3), SL(3,5), Spin(5,3). In each case one sees after the fact that the conjecture (and its converse) hold. A later calculation by the author (see (15.3) below) gives the Cartan 5-invariants of the twisted group SU(3,25), where the conjecture is again true; but sometimes $R_\lambda = Q_\lambda$ even when λ is irregular. (Subsequently, E. G. Zaslawsky [1] has computed the decomposition and Cartan matrices for a number of the small simple groups, including SL(3,5) = PSL(3,5), using known character tables.)

For SL(3,5), the relationship between R_λ and the Q_λ is quite systematic:

$$Q_{00} = R_{00} \oplus R_{04} \oplus R_{40} \oplus R_{44}$$
$$Q_{r0} = R_{r0} \oplus R_{r4} \quad \text{(for } r \neq 0\text{)};$$
$$Q_{0s} = R_{0s} \oplus R_{4s} \quad \text{(for } s \neq 0\text{)};$$
$$Q_{rs} = R_{rs} \quad \text{otherwise.}$$

If 4 is replaced by p-1, it can be shown (10.5) that the same pattern occurs for SL(3,p) in general. Note that this is at least compatible with the dimension comparison above and the known values of dim Q_{rs}. Indeed, $|\Gamma| = p^8 - p^3(p^3+p^2-1)$, so the second term measures the discrepancy $\Sigma(\dim M_\lambda)(\dim Q_\lambda - \dim R_\lambda)$. When λ is irregular, its dimension and p-dimension coincide (cf. (4.3)), so the decompositions indicated above agree with the numerical identity (for arbitrary integer p):

$$p^3 + p^2 - 1 = 5 + 2p(p+1)/2 + 6 \sum_{i=1}^{p-2} (i+1)(i+2)/2.$$

(d) Using his results on the "Brauer lifting", Lusztig [2, 5.6] proves that if $q \neq 2$, $M \otimes St_n$ is a PIM for $GL(n,q)$, where $q = p^n$ and where M is the usual n-dimensional representation of $GL(n,q)$. This adapts easily to $\Gamma_n = SL(n,q)$; the highest weight of M is the first fundamental weight λ_1, so when $n = 1$ the PIM in question is just $Q_\lambda = R_\lambda$, $\lambda = (p-1)\delta - \lambda_{n-1}$ ($p \neq 2$). Notice that λ is irregular precisely when $p = 2$.

(e) It has been noticed independently by the author and by Ballard that when $\lambda = 0$, R_λ is never equal to Q_λ. Here St occurs as a composition factor of Q_0 when viewed as a $K\Gamma$-module, and (being projective) must split off. To see this, note that the argument of (8.2) constructs Q_0 as a summand of $St \otimes St$, involving the highest weight $2(p-1)\delta$. But $M_{2(p-1)\delta} = M_{(p-2)\delta} \otimes M_\delta{}^{(p)}$. Upon restriction to Γ the right side involves a composition factor of highest weight $(p-2)\delta + \delta$ (7.2), (7.3).

(f) Sometimes one can force $R_\lambda = Q_\lambda$ in the regular case by comparison of dimensions. In his thesis, Ballard [1, §6] obtains a lower bound on dim R_λ by observing that R_λ is also a projective module for the "Borel subgroup" of Γ, whose PIM's are of dimension p^m and correspond bijectively to linear characters of the diagonal subgroup. (This parallels the arguments for $\underline{u}$ in Humphreys [1].) The conclusion is that dim $R_\lambda \geq |W\pi_\lambda| p^m$, where π_λ is the linear character of the diagonal subgroup of Γ corresponding to λ. Notice that the W-orbit of π_λ has the same cardinality as the W-orbit of λ in $X/(p-1)X$, which is the same as the cardinality of the W-orbit of $\lambda + \delta$ in $X/pX = \Lambda$. So the estimate becomes: dim $R_\lambda \geq a_\lambda p^m$ (a_λ = size of linkage class of λ).

In case Q_λ is of the "small" type considered in §9, with p not dividing f, we have dim $Q_\lambda = a_\lambda p^m$ and therefore $Q_\lambda = R_\lambda$. As a consequence, (9.2) yields the Brauer character of R_λ in this case. In general, Ballard's lower bound for dim R_λ is too small to be of help in proving Conjecture 10.3. But his results do help to reinforce the conjecture and to prove special cases of it.

10.4 Ballard's thesis

It may be helpful to outline briefly at this point some of the main ideas in Ballard [1].

His starting point is a construction of generalized characters of Γ_n (denoted by him G_σ), inspired by Srinivasan [3]. These characters ψ_μ are alternating sums of characters induced from linear characters of "Cartan subgroups" (groups of rational points of maximal tori of G defined over the finite field). It is shown that ψ_μ vanishes except at semisimple elements and can be expressed as an integer multiple of $s'(\mu) \cdot st$, where st is the Brauer character of St_n and where $s'(\mu)$ is the sum over the W-orbit of the (virtual) Brauer character associated with the character μ of T ($\mu \in X$). (Ballard denotes st by ϕ and $s'(\mu)$ by s_μ; this latter notation comes from Wong [1].)

Wong [1] had shown that the $s'(\mu)$ ($\mu \in X_{p^n}$) form a basis for the space of class functions on p-regular (i.e., semisimple) classes of Γ_n; another basis for this space consists of the Brauer characters ϕ_λ of the irreducible $K\Gamma_n$-modules M_λ ($\lambda \in X_{p^n}$). In a similar spirit, Ballard shows that the functions $s'(\mu) \cdot st$ form a basis for the space of projective Brauer characters; another basis for this space consists of the Brauer characters η_λ of the PIM's $R_{\lambda,n}$, denoted by him U(λ). In particular, st "divides" all principal indecomposable characters. The proof is based on a computation of the Brauer character of the module gotten by tensoring an irreducible module with St_n (cf. (9.2) above).

Let us give a more precise statement of Ballard's Theorem 2. Define $f(\mu) = (q-1)\delta + \sigma_o \mu$ in X, where $q = p^n$. Then for $\mu \in X_q$, $\eta_{f(\mu)}$ is the sum of $s'(\mu) \cdot st$ and integral multiples of various $s'(\nu) \cdot st$, where ν runs over weights in X_q strictly below μ in the partial order for which $f(\nu) > f(\mu)$. (Consequently, $R_{f(\mu),n}$ occurs exactly once as a summand of $\bar{V}_\mu \otimes St_n$ or $M_\mu \otimes St_n$.)

So far most of this is formal in nature. Ballard goes on to ask for conditions on μ which will insure that $s'(\mu) \cdot st$ is the Brauer character of a PIM. (Qualitatively, our approach indicates that this should be true mainly when μ is "small", i.e., the PIM is associated with the top alcove as in (9.2).) He defines a length function $\ln(\mu) = \langle \mu, \alpha_o \rangle$, where α_o is the highest short root, and shows that when $\ln(\mu) \leq q-1$, $s'(\mu) \cdot st$ is the Brauer character of the G-module (hence Γ_n-module) $\Sigma \bar{V}_{(q-1)+\sigma\mu}$ (using Weyl's character formula as in (9.2)).

In general, Ballard obtains a lower bound for the dimension of a PIM; this is based , as already mentioned, on a comparison with PIM's of a Borel subgroup. He also gets an upper bound on the dimension, by looking at the relevant linkage class component of a tensor product with St_n. In particular, he obtains by a somewhat complicated argument a result equivalent to our Theorem 10.2, but without an explicit description of Q_λ or its dimension except in the special situation of (9.2).

10.5 An example: SL(3,p)

Let $\Gamma = SL(3,p)$. We leave aside the case $p = 3$, which is already treated directly in Humphreys [5]. According to (10.4), part (f), $R_\lambda = Q_\lambda$ when $d_\lambda = 1$ (e.g., when λ lies in the top alcove). To verify the relationship between arbitrary R_λ and Q_λ asserted in (10.3), it suffices according to the final remarks there to show either that the Q_λ decompose for Γ at least as much as claimed or that they decompose at most as much as claimed. Let us sketch an argument for the former, based on comparison of Brauer characters. Suppose λ is irregular and lies in the (interior of the) bottom alcove, so that $d_\lambda = 2$ and dim $Q_\lambda = 12p^3$. As in (8.2), Q_λ occurs once as a summand of $M_\mu \otimes St$, where $\mu = (p-1)\delta + \sigma_o \lambda$ in X^+. The only lower dominant weight ν of M_μ for which $\nu + (p-1)\delta$ is linked to λ is easily seen to be $\lambda + \delta$. So the calculation in (9.2) shows that the formal character of the direct summand of the tensor product belonging to this linkage class is $s(\mu)\cdot st + ks(\lambda + \delta)\cdot st$ (k = multiplicity of $\lambda + \delta$ in M_μ). This can also be interpreted as the Brauer character for Γ. Since λ is irregular, μ involves a coordinate $p-1$, so that $s(\mu)$ has formal degree greater than that of $s'(\mu)$, usually 6 as opposed to 3. On the other hand, $s(\lambda+\delta) = s'(\lambda+\delta)$ as Brauer characters, and $s(\lambda+\delta)\cdot st$ is the Brauer character of $Q_{\lambda^o} = R_{\lambda^o}$.

These considerations force us to conclude that Q_λ has formal character $s(\mu)\cdot st + s(\lambda+\delta)\cdot st$. On the other hand, Ballard [1, Thm. 2, p. 38] shows that the Brauer character of R_λ is $s'(\mu)\cdot st$ plus possibly some multiple of $s'(\lambda+\delta)\cdot st = s(\lambda+\delta)\cdot st$. Comparison of dimensions implies that Q_λ decomposes for Γ at least as much as claimed. The boundary cases with $d_\lambda = 2$ are treated similarly.

Let us conclude by considering Cartan invariants.

In (9.3) we determine the G-composition factors of Q_λ when λ lies in the top alcove; then $Q_\lambda = R_\lambda$. The twisted tensor products for G become ordinary tensor products for Γ (7.3), whose composition factors can be read off rather easily (7.2). In particular, there are usually 54 of these, 18 belonging to big weights (in the top alcove) and 36 to small weights (in the bottom alcove). The generic pattern is quite regular and leads to generic Cartan invariants for PIM's of dimension $6p^3$:

big μ	6	2	2	2	2	2	2						
small μ^o	6	4	4	4	4	4	4	1	1	1	1	1	1

Here λ and λ° each occur six times, while other weights μ, μ° occur as indicated. Since C is symmetric (R.3), it is not too hard to deduce the generic nonzero Cartan invariants for PIM's of dimension $12p^3$:

```
 6  4  4  4  4  4  4  1  1  1  1  1  1
12  6  6  6  6  6  6  2  2  2  2  2  2  2  2  2  2  2  2
```

We remark that for types other than A_2, we may again expect to find generic Cartan invariants for PIM's of each generic dimension. These have not yet been worked out in detail, however.

POSTSCRIPT (April 1976)

J.C. Jantzen has pointed out that the assertion of page 42, lines 3-4, requires further explanation. This point will be discussed in detail elsewhere, in conjunction with remark (d) of (8.3). It does not affect the special case treated in (9.2).

III. ORDINARY REPRESENTATIONS

In this part we shall consider the relationship between the ordinary and the modular representations of Γ.

As a first reduction, consider the blocks of $K\Gamma$ (R.1), each of which involves a certain subset of the R_λ (or M_λ) and a certain subset of the $\bar{Z}_i$. It is well known that $K\Gamma$ has a unique block of defect 0, involving just $St = R_{(p-1)\delta} = M_{(p-1)\delta}$ (= the reduction modulo p of a $\mathbb{C}\Gamma$-module of dimension p^m), cf. (10.1). At the other extreme, it is shown in Humphreys [2] (cf. Dagger [2]) that there are as many blocks of highest defect as the order of the center of Γ, and that there are no further blocks. The center of Γ has order dividing $f = [X:X_r]$: For type D_ℓ (ℓ even) the center has order 4 (resp. 1) when p is odd (resp. p = 2). For other types, the order is $d = g.c.d.$ (f, p-1). In this latter case, it is easy to see that we have the following distribution of the M_λ among the blocks of highest defect: X/X_r has a unique subgroup of index d, Y/X_r. Take the restricted weights (other than $(p-1)\delta$) in a single coset of Y in X, to get the M_λ occurring in a single block. (Compare the various quotients of Γ by subgroups of its center.)

§11. The Brauer tree of SL(2,p)

Here we shall review the well known case $\Gamma = SL(2,p)$ (with p assumed to be odd unless otherwise stated). Our intention is to reformulate the "Brauer tree" in a way which emphasizes the geometry of the affine Weyl group. Relevant references include Brauer, Nesbitt [1], Dornhoff [1, Part B, §71], Srinivasan [1], Jeyakumar [1], Humphreys [4, 7].

11.1 Modular representations

First we recall what is known about various $K\Gamma$-modules. Weights
are identified with rational integers, the restricted weights
$\Lambda = \{0,1,\ldots,p-1\}$ being sorted into linkage classes $\{0,p-2\}$, $\{1,p-3\}$,
$\ldots$, $\{(p-3)/2,(p-1)/2\}$, $\{p-1\}$. The irreducible module M_λ has dimension
$\lambda + 1$, while $Z_\lambda \underset{\underline{u}}{\leftrightarrow} M_\lambda \oplus M_{p-2-\lambda}$ (except when $\lambda = p-1$, $M_\lambda = Z_\lambda = St$). Thus
$\dim Q_\lambda = 2p$ unless $\lambda = p-1$. Viewed as a G-module, $Q_\lambda \underset{G}{\leftrightarrow} M_\lambda \oplus M_{2p-2-\lambda}$
$\oplus\, M_\lambda$ ($\lambda \neq p-1$), where $M_{2p-2-\lambda} = M_{p-2-\lambda} \otimes M_1^{(p)}$. This is easily deduced
from the tensor product construction, cf. (9.3). In turn,
$Q_\lambda \underset{K\Gamma}{\leftrightarrow} M_\lambda \oplus (M_{p-2-\lambda} \otimes M_1) \oplus M_\lambda$, cf. Proposition 7.3(a). In view of (7.2),
$M_{p-2-\lambda} \otimes M_1 \underset{K\Gamma}{\leftrightarrow} M_{p-1-\lambda} \oplus M_{p-3-\lambda}$ provided $p-3-\lambda \geq 0$, while if $\lambda = p-2$, this
tensor product is just M_1 ($= M_{p-1-\lambda}$). As recalled in (10.3),
$Q_\lambda = R_\lambda$ ($\lambda \neq 0$), $Q_0 = R_0 \oplus St$.

The two pairs of weights $\{\lambda,p-1-\lambda\}$, $\{\lambda,p-3-\lambda\}$ which usually
figure in the $K\Gamma$-module Q_λ may be thought of as "deformations" of the
linkage class $\{\lambda,p-2-\lambda\}$ which figures in the $\underline{u}$-module Q_λ. Notice that
the dimensions of the M_μ corresponding to each pair add up to $p+1$
(resp. $p-1$), whereas the linkage class yields $\dim Z_\lambda = p$.

11.2 The Brauer tree

It is well known that $\mathbb{C}\Gamma$ has $p+4$ distinct
irreducible modules Z_i (p being odd), of which $(p-3)/2$ have dimension
$p+1$, $(p-1)/2$ have dimension $p-1$, and the remaining ones have
respective dimensions 1, p, $(p+1)/2$, $(p+1)/2$, $(p-1)/2$, $(p-1)/2$. This
information is easily obtainable from the modular theory, as follows,
using the basic reciprocity rules (R.4). Denote by $\bar{Z}_i$ some reduction
modulo p of Z_i.

Consider a "typical" case $Q_\lambda = R_\lambda$ (excluding $\lambda = 0$, $p-1$, $p-2$,
$(p-1)/2$, $(p-3)/2$). There are three distinct $K\Gamma$-composition factors
M_λ, $M_{p-1-\lambda}$, $M_{p-3-\lambda}$, with M_λ repeated twice. Reciprocity makes it clear
that R_λ involves exactly two distinct $\bar{Z}_i$, each having M_λ as a

composition factor. How the other two M_μ are allotted becomes clear
if we compare the "neighboring" PIM's, since the composition factors
shared by two PIM's must be those of one or more $\bar{Z}_i$. We can picture
the PIM's as follows:

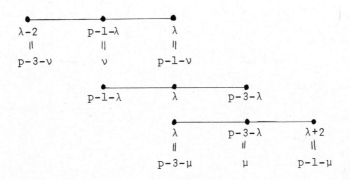

Here the middle vertex is to be viewed as having multiplicity 2. The
conclusion is that each $\bar{Z}_i$ occurring here has two composition factors,
corresponding to the deformations of the linkage class $\{\lambda,\, p-2-\lambda\}$
discussed in (11.1). The atypical cases can be analyzed similarly,
e.g., R_{p-2} involves a $\bar{Z}_i$ of dimension p+1 whose composition factors
are M_{p-2}, M_1, along with a $\bar{Z}_i$ of dimension p-1 having M_{p-2} as its sole
composition factor.

It is convenient to integrate all PIM's into a single graph
having two connected components (one for each block of highest defect),
the Steinberg module being left aside. One merely has to superimpose
all the pictures above. For example, when p = 11 we obtain:

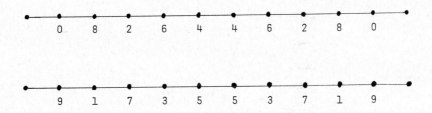

Notice that all vertices are repeated. The middle edge involves two $\bar{Z}_i$ of equal dimension, either $(p-1)/2$ or $(p+1)/2$, whereas the other edges correspond to single modules $\bar{Z}_i$. If numbering of vertices were continued, we would get (respectively) $p-1$, -1.

We have obtained for each block a graph which is a <u>tree</u> (connected and having no circuits). The fact that the incidence relation between the R_λ and the $\bar{Z}_i$ leads to a tree was appreciated by Brauer; it appears as a special case of the general theory of blocks with cyclic defect group, developed by Brauer and Dade (cf. Dornhoff [<u>1</u>, Part B, §70]). However, in this theory the <u>vertices</u> correspond to the $\bar{Z}_i$ and the <u>edges</u> to the M_λ. It happens that for SL(2,p) the tree can be formulated either way. In Brauer's version, the pair of $\bar{Z}_i$ having dimension $(p+1)/2$ or $(p-1)/2$ share an end vertex; in our version, they share a middle edge.

Our tree for (say) the even block should be thought of as a model of the restricted weight region, with only the even weights labelled:

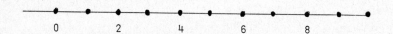

An unlabelled vertex can be viewed as a "wall" of the two neighboring "chambers" (edges); reflection across this wall adds or subtracts the root $\alpha_1 (= 2\lambda_1)$. To label these vertices correctly, use the recipe: Replace an odd weight by the even weight linked to it.

If $p = 2$, there is just one block of highest defect, and the tree looks like:

11.3 Comparison of Brauer characters

It is also possible to work out the relationship between ordinary and modular representations of Γ in terms of characters. The character table of SL(2,p) is reproduced below, cf. Schur [1], Dornhoff [1, Part A, §38], Humphreys [7]. Representatives of conjugacy classes (except those of zc, zd) are listed across the top.

	1	z	a^ℓ	b^m	c	d
1_G	1	1	1	1	1	1
ψ	p	p	1	-1	0	0
ζ_i	p+1	$(-1)^i(p+1)$	$\tau^{i\ell}+\tau^{-i\ell}$	0	1	1
ξ_1	$\frac{1}{2}(p+1)$	$\frac{1}{2}\varepsilon(p+1)$	$(-1)^\ell$	0	$\frac{1}{2}(1+\sqrt{\varepsilon p})$	$\frac{1}{2}(1-\sqrt{\varepsilon p})$
ξ_2	$\frac{1}{2}(p+1)$	$\frac{1}{2}\varepsilon(p+1)$	$(-1)^\ell$	0	$\frac{1}{2}(1-\sqrt{\varepsilon p})$	$\frac{1}{2}(1+\sqrt{\varepsilon p})$
θ_j	p-1	$(-1)^j(p-1)$	0	$-(\sigma^{jm}+\sigma^{-jm})$	-1	-1
η_1	$\frac{1}{2}(p-1)$	$-\frac{1}{2}\varepsilon(p-1)$	0	$(-1)^{m+1}$	$\frac{1}{2}(-1+\sqrt{\varepsilon p})$	$\frac{1}{2}(-1-\sqrt{\varepsilon p})$
η_2	$\frac{1}{2}(p-1)$	$-\frac{1}{2}\varepsilon(p-1)$	0	$(-1)^{m+1}$	$\frac{1}{2}(-1-\sqrt{\varepsilon p})$	$\frac{1}{2}(-1+\sqrt{\varepsilon p})$

Characters of SL(2,p) (classes of zc, zd omitted)

Here

$$z = \begin{pmatrix} -1 & 0 \\ 0 & -1 \end{pmatrix}, \quad c = \begin{pmatrix} 1 & 1 \\ 0 & 1 \end{pmatrix}, \quad d = \begin{pmatrix} 1 & \nu \\ 0 & 1 \end{pmatrix}, \quad a = \begin{pmatrix} \nu & 0 \\ 0 & \nu^{-1} \end{pmatrix},$$

ν being a generator of the multiplicative group of $\mathbb{F}_p$, and b denotes an element of order p+1 which is diagonalizable over a quadratic extension of $\mathbb{F}_p$. The irreducible characters are listed along the left side. In the table, τ is a primitive $(p-1)^{st}$ root of 1, σ is a primitive $(p+1)^{st}$ root of 1, $\varepsilon = (-1)^{(p-1)/2}$. The various indices range as follows: $1 \le \ell \le (p-3)/2$, $1 \le m \le (p-1)/2$, $1 \le i \le (p-3)/2$, $1 \le j \le (p-1)/2$.

The Brauer characters of the irreducible modular representations are not too difficult to write down (cf. Srinivasan [1]), and comparison with the ordinary characters then yields the Brauer characters of PIM's (principal indecomposable characters). For this purpose, one of course ignores the last two columns of the character table as well as the columns which correspond to the omitted classes.

Alternatively, we can write down the principal indecomposable characters using Theorem 9.2 and the remarks in (10.3); this requires knowing the values of the Steinberg character ψ. The way in which ordinary characters combine to form principal indecomposable characters is then rather transparent.

For groups other than $SL(2,p)$, we could also use the principal indecomposable characters to good advantage if we knew them a priori as we do here. But our present information is precise only for the "small" PIM's as in (9.2).

11.4 The case $SL(2,p^n)$

Let us consider briefly what happens when Γ is replaced by $\Gamma_n = SL(2,p^n)$. Set $q = p^n$. Here there are known to be $q+4$ irreducible modules over $\mathbb{C}$, roughly half of dimension $q+1$ and half of dimension $q-1$. Using the known characters (resp. Brauer characters), Srinivasan [1] determined the composition factors of the $\overline{Z}_i$. As n gets large, the number of these also gets large, thereby making it difficult to picture the incidence relation by a graph. In fact, there are usually 2^n composition factors. But these can be organized systematically into two bunches, corresponding to the members of a deformed linkage class. For example, let $q = p^2$, and denote by (r_0, r_1) the weight $r_0 + pr_1$. Let $\phi(r_0, r_1)$ be the Brauer character attached to $M_{r_0} \otimes M_{r_1}(p)$. Set $\Phi(r_0, r_1) = \phi(r_0, r_1) + \phi(p-2-r_0, r_1-1)$

Then $\Phi(r_0,r_1) + \Phi(p-1-r_0,p-1-r_1)$ is the Brauer character of some $\bar{Z}_i$ of dimension $p^2 + 1 = (\lambda+1) + (p^2-\lambda)$, $\lambda = r_0 + pr_1$.

§12.   The Brauer complex of SL(3,p)

For a group of rank $\ell > 1$, a graph is no longer adequate to depict accurately the incidence relation involving the $\bar{Z}_i$ and the R_λ (or M_λ). We propose instead, for each block of highest defect in $K\Gamma$, an ℓ-dimensional chamber complex, which looks like the fundamental domain (alcove) of W_a filled with p^ℓ small copies of itself. This is of course meant to generalize the Brauer tree of SL(2,p), as reformulated in §11, and will be called the Brauer complex. In this section we give a fairly detailed description in case $\Gamma = $ SL(3,p).

12.1   Deformations of linkage classes

Let us first review what is known about modular representations, with emphasis on the generic situation. Call $\lambda \in X_p$ (or Λ) big (resp. small) if it lies in the interior of the top (resp. bottom) alcove for W_a; then the linkage class of λ has cardinality 6, unless $p = 3$ (a case which we temporarily leave aside). Recall that $\bar{V}_\lambda = M_\lambda$ if λ is small, while $\bar{V}_\lambda \overset{\leftrightarrow}{G} M_\lambda \oplus M_{\lambda^o}$ if λ is big. Also, Z_λ has (in general) 9 composition factors, the 3 small weights linked to λ in Λ each occurring twice.

When λ is big, the composition factors of Q_λ (as G-module or as $K\Gamma$-module) can be read off from (9.3), (10.5), and §7. Here $Q_\lambda = R_\lambda$ has dimension $6p^3$. If all weights linked to λ are sufficiently far from the walls of their alcoves and from each other, there will be 54 $K\Gamma$-composition factors, belonging to 18 big and 36 small weights as indicated in (10.5). In particular, λ will occur 6 times, which already implies (thanks to (R.4)) that R_λ can involve at most 6 of the $\bar{Z}_i$. It still has to be asked how the composition factors of R_λ are distributed among the various $\bar{Z}_i$. We claim that a simple pattern governs this, as in the case of $SL(2,p)$.

In view of (9.3), (7.2), (7.3), the $K\Gamma$-composition factors of Q_λ (= R_λ here) have highest weights "close to" the weights in Λ linked to λ. So we think in terms of deforming the linkage class. Let $\lambda(\sigma)$ (= $\sigma(\lambda+\delta) - \delta$ in Λ) be the weight linked to λ by σ, and define δ_σ as in (6.1) to be the sum of those λ_i for which $\ell(\sigma_i\sigma) < \ell(\sigma)$, cf. Table 1 below. For each $\tau \in W$, we deform the linkage class slightly by adding $\sigma\tau\sigma^{-1}(\delta_\sigma)$ to $\lambda(\sigma)$. The weights $\sigma\tau\sigma^{-1}(\delta_\sigma)$ appear in Table 2, the rows there being indexed by σ and the columns by τ.

TABLE 1

σ	linked weight	δ_σ
1	(r,s)	(0,0)
σ_1	(p-r-2,r+s+1-p)	(1,0)
σ_2	(r+s+1-p,p-s-2)	(0,1)
$\sigma_1\sigma_2$	(2p-r-s-3,r)	(1,0)
$\sigma_2\sigma_1$	(s,2p-r-s-3)	(0,1)
σ_o	(p-s-2,p-r-2)	(1,1)

TABLE 2

	1	σ_o	σ_1	σ_2	$\sigma_1\sigma_2$	$\sigma_2\sigma_1$
1	(0,0)	(0, 0)	(0, 0)	(0, 0)	(0, 0)	(0, 0)
σ_1	(1,0)	(1, 0)	(-1,1)	(0,-1)	(0,-1)	(-1,1)
σ_2	(0,1)	(0, 1)	(-1,0)	(1,-1)	(1,-1)	(-1,0)
$\sigma_1\sigma_2$	(1,0)	(-1,1)	(1, 0)	(0,-1)	(-1,1)	(0,-1)
$\sigma_2\sigma_1$	(0,1)	(1,-1)	(-1,0)	(0, 1)	(-1,0)	(1,-1)
σ_o	(1,1)	(-1,-1)	(2,-1)	(-1,2)	(1,-2)	(-2,1)

As τ runs over W, the deformed weights run over (generically) 36 of the 54 μ for which M_μ occurs as a composition factor of R_λ, the other 18 being the small weights below and linked to the big weights in this list. We refer to the τ-<u>deformation</u> associated with λ as the corresponding set of 9 weights, 3 big and 6 small. Schematically, each of the following configurations occurs 3 times:

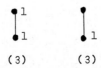

THEOREM. Fix τ ∈ W. Then Σp-dim(μ), sum over the τ-deformation associated with any big weight λ, equals:

(A) $p^3 + 2p^2 + 2p + 1$ if τ = 1,

(B) $p^3 - 1$ if τ = $\sigma_1, \sigma_2, \sigma_0$,

(C) $p^3 - p^2 - p + 1$ if τ = $\sigma_1\sigma_2$, $\sigma_2\sigma_1$.

This is stated in Humphreys [6]. The proof is a simple direct computation: one has only to add 6 Weyl dimensions dim(μ). Even if λ lies near a linked weight or near a wall of the top alcove, the same formal calculation works, although some weights may be repeated or may lie outside the restricted region; similarly if p = 3. (Weyl's dimension polynomial can be applied to any weight.) For example, let p = 5 and let λ = (3,3), so its linkage class is given by Table 1 as follows: (3,3), (0,2), (2,0), (1,3), (3,1), (0,0). If τ = $\sigma_1\sigma_2$, the τ-deformation is therefore: (3,3) along with (0,0); (0,1); (3,-1); (0,4) along with (-1,3); (2,1); (1,-2). The weight (2,1) is on the wall between the top and bottom alcove, so it is its own opposite linked weight (and is only counted once). The Weyl dimension of (3,-1) or (-1,3) being 0, these weights can essentially be ignored. But (1,-2) has Weyl dimension -1, balancing the contribution of (0,0) to the polynomial $p^3 - p^2 - p + 1 = 96$.

The three polynomials in the theorem are known (see, for example, Simpson and Frame [1]) to be the "generic" degrees of the three large families of irreducible representations of $C\Gamma$, each family occurring with approximately the frequency 1/6, 1/2, 1/3 of the indicated conjugacy class of W. It is difficult to resist the suspicion that we have found the composition factors of the corresponding $\bar{Z}_i$. (For the first of these families, this has been discovered independently by Carter, Lusztig [2].) However, our formal deformation patterns must be interpreted with some care when weights occur whose associated Weyl dimensions are negative; then some other weight must be cancelled from the list. This shows up in the above example for p = 5, where (0,0) will not appear as highest weight of a composition factor for some $\bar{Z}_i$ involving (3,3); this contributes in turn to the fact that $R_{(0,0)}$ is smaller than $Q_{(0,0)}$.

As we suggested in the Introduction, there is an underlying analogy between the $\underline{u}$-modules Z_λ and the $K\Gamma$-modules $\bar{Z}_i$. The formal identities in the theorem should therefore be compared with the identity: $p^3 = \Sigma d_\mu p\text{-}dim(\mu)$, sum over the linkage class of λ.

12.2 Picturing PIM's

In order to relate the formal deformation patterns of (12.1) to the actual behavior of the $\bar{Z}_i$, it is necessary as in §11 to examine how the different PIM's overlap. For this a picture is again helpful. Starting with a big weight $\lambda = r\lambda_1 + s\lambda_2 = (r,s)$, construct a regular hexagon with λ as center and divide this into six small alcoves as indicated in the example below. Label the vertices of the hexagon by the other 6 (in general distinct) big weights occurring in the various deformation patterns (each of these in general figures twice in $Q_\lambda = R_\lambda$). An alcove is labelled τ if its vertices are the three big weights occurring in the τ-deformation associated with λ. The ordering of vertices is chosen to make the hexagon resemble the

"Coxeter complex" of W: the lines through λ are to be thought of as the reflecting hyperplanes for the various σ_α. The example shown involves $p = 5$, $\lambda = (3,2)$.

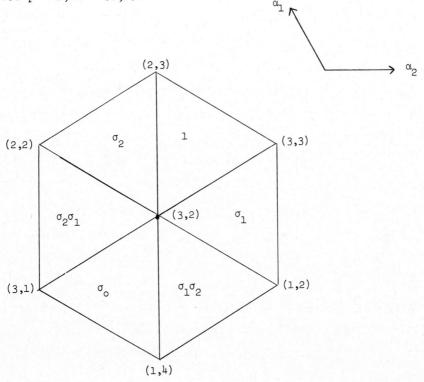

Only the big weights figuring in R_λ have been pictured explicitly. By symmetry (R.3), M_μ occurs as a composition factor of R_λ just as often as M_λ occurs in R_μ. So it is enough to provide a picture of the PIM belonging to a small weight μ, say λ°, which generically has dimension $12p^3$. For this we add to the hexagon for λ six small alcoves, to form a star-shaped figure as indicated below. The new vertices are labelled with the big weights gotten by subtracting from λ each of the 6 roots (these occur in the bottom row of Table 2). For example, when $p = 5$, we go beyond the alcove labelled σ_1 by subtracting from $(3,2)$ the root $\alpha_2 = (-1,2)$, to get $(4,0)$. (The outer wall of this alcove may be thought of as a reflecting hyperplane for σ_2.) In this example, some of the new weights obtained in this way will be irregular or even non-restricted, which shows up in the fact that $Q_{(1,0)}$ $\neq R_{(1,0)}$. The rationale for our labelling of vertices will become clearer below.

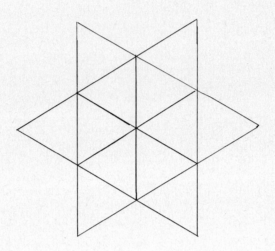

12.3 The Brauer complex

Our objective is to integrate all the pictures of PIM's just described into a single complex (for each block of highest defect in $K\Gamma$). The number of blocks involved is the g.c.d. of p-1 and $3 = [X:X_r]$, i.e. 1 if $p \not\equiv 1 \pmod 3$, 3 otherwise. (The cases p = 5,7 illustrated below are typical, though not large enough to exemplify the "generic" situation in which a small PIM has a total of 54 composition factors.) In the case of 3 such blocks, the weights involved in a block lie in just one of the cosets of X/X_r.

Begin with the top alcove (for W_a) meeting X_p, whose lattice points are labelled with the big weights (we also consider weights in the closure of this alcove). Pass through each lattice point lines (= hyperplanes) parallel to the walls of the alcove, thereby

SL(3,5)

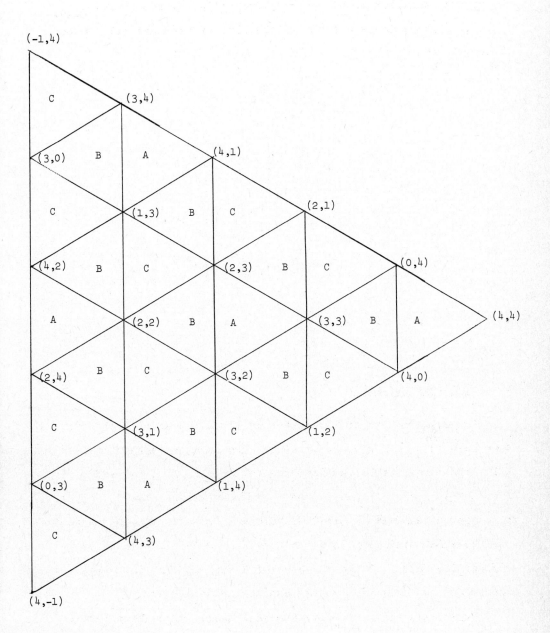

SL(3,7)

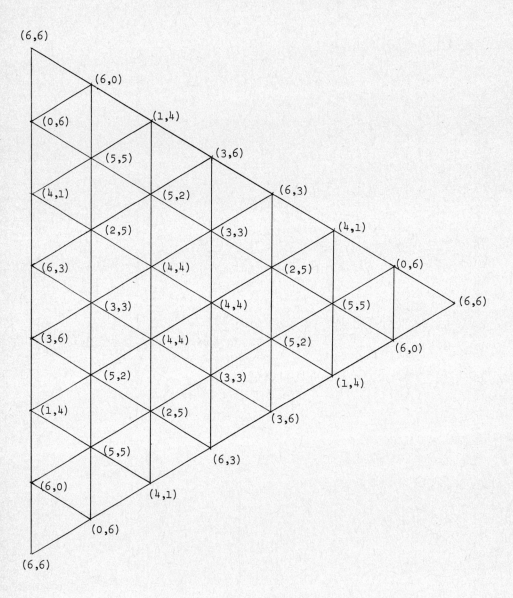

SL(3,7)

(reverse coordinates to get
third block)

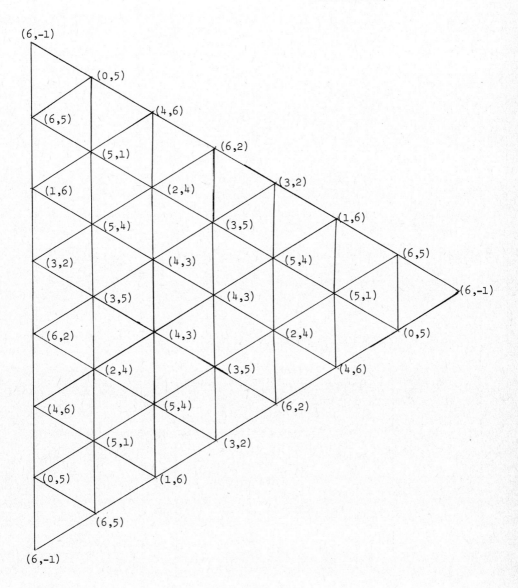

subdividing the alcove into p^2 small copies of itself. Choose a coset of X/X_r containing weights which correspond to the block of $K\Gamma$ in question; if $p \equiv 2 \pmod 3$, any coset will do, e.g., the one containing $(p-1)\delta$. Let $W^1 = \{1, \sigma_1\sigma_2, \sigma_2\sigma_1\}$; this is the subgroup of W which links a weight in the top alcove to another weight in the top alcove. Notice that for each weight λ, there is a unique $\sigma \in W^1$ for which $\lambda - \delta_\sigma$ belongs to the chosen coset of X/X_r, in view of the fact that $(0,0)$, $(1,0)$, $(0,1)$ represent the three cosets. Given a big weight λ, we replace it by the linked weight $\lambda(\sigma)$ to obtain the desired labelling of the vertices of our complex. More precisely, we replace λ by the indicated linked weight in Table 1, even if the weights involved lie outside the restricted weight region. For example, when $p = 5$, the weight $(-1,4)$ on the boundary may be replaced by $(4,-1)$ if the chosen coset is that of $(4,4)$. In cases like this, it should be clear that the choice of a different coset would merely re-orient the complex.

It has to be shown that this labelling agrees with the description of a single PIM in (12.2). To follow the argument, the reader may find it helpful to compare the example given there ($p = 5$, $\lambda = (3,2)$) with the original configuration of weights pictured below; we are using the coset of $(4,4)$ and therefore leave the weights $(3,3)$, $(2,2)$, $(1,4)$ in their original positions. Consider the diagram in (12.2) first, with λ denoting the weight in the position of $(3,2)$ and μ the weight in the position of $(3,3)$. According to the recipe, $\mu = \lambda(\sigma) + \sigma\tau\sigma^{-1}(\delta_\sigma)$ (equality in Λ), where σ lies in W^1; here $\sigma = \sigma_1\sigma_2$ and $\tau = 1$. But originally the two weights were $\mu = (3,3)$ and $\mu - \delta_\sigma = (2,3)$. Subtracting $\delta_{\sigma^{-1}} = (0,1)$ from $(2,3)$ gets us into the chosen coset, so our new recipe tells us to replace $(2,3)$ by the linked weight $\sigma^{-1}(\mu - \delta_\sigma + \delta) - \delta$ (computed in Λ); but this is precisely λ. The weights in other positions can be analyzed similarly.

Consider next our picture in (12.2) of a PIM of (generic) dimension $12p^3$. Arguments similar to the one just sketched show that

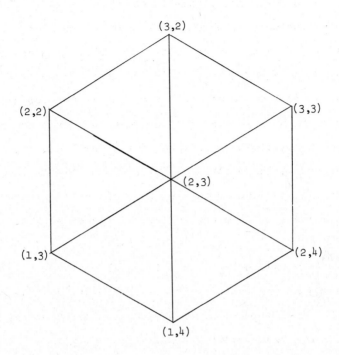

that picture is compatible with the complex just defined. We shall
not belabor this point; but it may be well to emphasize how easily the
deformation patterns of (12.1) can be read off. Start with $p = 5$,
$\lambda = (3,2)$, and select τ. The big weights involved in the τ-deformation
associated with (3,2) are the vertices of the alcove labelled τ in
(12.2), and with each we take its opposite linked weight. The remaining three small weights are found by "reflecting" the three vertices
of the small alcove in question across the opposite walls and passing
to opposite linked weights. For example, when $\tau = 1$, we start with
(3,2), (2,3), (3,3) and the linked weights (1,0), (0,1), (0,0); we
then "reflect" to get (2,1), (1,2), (2,2), and replace these by the
linked weights (2,1), (1,2), (1,1).

Notice that the weights lying on a wall of the complex are the
weights belonging to the block in question which were originally in a
wall of the top alcove. This shows up clearly in the example just
mentioned.

In picturing a single PIM of dimension $6p^3$, we labelled each small alcove with an element of W. If another PIM shares this alcove, it may or may not be labelled with the same element of W, but the choice varies within a <u>conjugacy class</u> of W. For example, when p = 5, the alcove labelled σ_1 for λ = (3,2) is the same as the alcove labelled σ_2 for (3,3). However, the labels do not change in case τ = 1, $\sigma_1\sigma_2$, or $\sigma_2\sigma_1$ (as the reader can verify). We have simply labelled alcoves A,B,C, to indicate the polynomials associated with deformation patterns in Theorem 12.1.

At this point the Brauer complex of a block just provides a convenient pictorial summary of the information we have about PIM's. The next task is to show that the deformation patterns faithfully portray the $\bar{Z}_i$ (at any rate, those having the generic degrees in (12.1)). As in §11, it is necessary to consider how PIM's overlap, sharing the composition factors of one or more $\bar{Z}_i$'s. It is clear that three hexagons will share at most one small alcove. Indeed, an enumeration of composition factors shows that the big weights shared by three overlapping PIM's of dimension $6p^3$ will belong to the vertices of their common alcove, each weight occurring twice. In the generic situation, this further implies that each such PIM involves at least (hence exactly) 6 distinct $\bar{Z}_i$, hence that the big weights of such PIM's occur just once in each $\bar{Z}_i$. So in fact there is just one $\bar{Z}_i$ common to all three PIM's. Reciprocity shows that its composition factors are precisely those occurring in the deformation pattern associated with this alcove.

It must be emphasized that this sort of analysis is valid only in the generic situation; degeneracies will be discussed in (12.4).

Since the Brauer characters of PIM's are known in detail for SL(3,p), cf. (10.5), it would also be possible in principle to compare these with the known characters of Γ (cf. Simpson, Frame [1]). For this purpose it would be preferable to replace the big weight

$\lambda = (r,s)$ by the small weight $\mu = (p-1-s, p-1-r)$ used in the tensor product construction of (8.2), since the virtual Brauer character $s(\mu)$ figures directly in the Brauer character of R_λ. In the generic case, the $\bar{Z}_i$ occurring in R_λ are those for which the ordinary character arises (cf. Deligne, Lusztig [1]) from a character of a torus defined by the weight μ. The reader may find it helpful to compare our Brauer complex for $p = 5$ with the character table of SL(3,5), in order to appreciate how the Brauer characters behave. (Cf. also the table of Cartan invariants in Humphreys [5].)

12.4 Degeneracies

When $p \equiv 1 \pmod 3$, there are 3 blocks of highest defect. In the Brauer complex of a block, every vertex is then repeated 3 times, and as a result the small alcove at the center of the complex has all its vertices labelled with the same big weight. It follows (e.g., for reasons of reciprocity) that there are three $\bar{Z}_i$ here, of equal degree and having the same composition factors; indeed, the Brauer characters of these $\bar{Z}_i$ must be the same. They will be of type A or C depending on whether the block is that of the 1-character 1_Γ or not. The corresponding ordinary characters appear in columns 7, 10, 11 of the character table in Simpson, Frame [1]. When there is only one block, no such phenomenon occurs. Notice too that in the case of 3 blocks, the $\bar{Z}_i$ in general position occur 3 times in the complex.

The boundary behavior is also interesting, and helps to illuminate the regularity criterion (10.5): $Q_\lambda = R_\lambda$ if and only if λ is regular. In the interior, except as noted in the preceding paragraph, each PIM of dimension $6p^3$ involves exactly six $\bar{Z}_i$ of type, frequency, and degree indicated by Theorem 12.1. But an alcove of type A located at the boundary (though not at the intersection of two outer walls) involves two $\bar{Z}_i$, of respective degrees $p^3 + p^2 + p$, $p^2 + p + 1$. Notice that if we had continued the complex across the outer wall we would reach a neighboring alcove of type B. The degrees involved are $p^3 + 2p^2 + 2p + 1$, $p^3 - 1$, and these can be combined formally to yield the the degenerate degrees above, by taking half their sum (resp. half their difference). The coefficients here are taken from the rows of the character table

$$
\begin{array}{c|cc}
 & 1 & 1 \\
 & 1 & -1
\end{array}
$$

of the subgroup of W of order 2 generated by the reflection in the wall in question, but this is not yet fully explained. We remark that the larger degenerate $\bar{Z}_i$ has a repeated composition factor of highest weight λ^o, while this composition factor is absent from the smaller degenerate $\bar{Z}_i$. This helps to explain the behavior of the PIM's in these cases, e.g., when $p = 7$, $R_{(1,3)}$ has dimension $12p^3$ but involves two occurrences of a $\bar{Z}_i$ of dimension $p^3 + p^2 + p$ rather than one occurrence each of generic $\bar{Z}_i$ of type A,B (one point of the "star" is cut off by a wall of the complex).

Where two walls of the complex meet, even more degeneracy is possible; this is governed in some way by the character table of the subgroup of W generated by the reflections in these two walls (i.e., W itself):

$$
\begin{array}{c|ccc}
 & 1 & 1 & 1 \\
 & 1 & -1 & 1 \\
 & 2 & 0 & -1
\end{array}
$$

Notice what happens if we start with the six $\bar{Z}_i$ in a typical PIM $Q_\lambda = R_\lambda$ of dimension $6p^3$ (λ big), constructed inside $M_\mu \otimes St$, and formally combine their Brauer characters using the rows of this character table. The dimensions add up as follows:

$$A + 3B + 2C = 6p^3$$
$$A - 3B + 2C = 6$$
$$2A \qquad - 2C = 6(p^2+p).$$

The Brauer characters themselves add up to (respectively):
$s(\mu) \cdot st$, $s(\mu) \cdot 1_\Gamma$, $s(\mu) \cdot \chi$, where st = Steinberg character, 1_Γ, and χ
are the distinct constitu ents of the character of Γ induced from the
1-character of a Borel subgroup of Γ. Here $s(\mu)$ denotes the formal
sum over the W-orbit of μ, interpreted as a Brauer character (cf.
(10.5)). It is instructive to compare identity (5) in Solomon [1],
which formally resembles the preceding calculation. (Here the
irreducible characters of W happen to be compounds of the reflection
character, so Solomon's identity covers the situation nicely; for
other types of Weyl group, other identities may be required.)

Returning to the Brauer complex, we note that the top alcove in
the block involving (0,0) will contain 1_Γ, χ, st (which strictly
speaking goes into its own block). If there are other blocks, however,
the top alcove will correspond to a single $\bar{Z}_i$ of type C, with no
degeneracy involved.

It is interesting to examine the PIM $R_{(0,0)}$ in detail. There are
7 alcoves involved, the 6 surrounding the weight (p-2,p-2) and the 1
having vertex (p-1,p-1). Among the former, the alcove labelled A
gives rise to 2 degenerate $\bar{Z}_i$ of dimension $p^3 + p^2 + p$ having (0,0) as
highest weight of a composition factor, rather than a single $\bar{Z}_i$ of the
generic dimension (cf. the discussion above). Formally, we have the
dimension equation: $7p^3 = 1 + 2(p^3 + p^2 + p) + 3(p^3 - 1) + 2(p^3 - p^2 - p + 1)$.

§13. The Brauer complex of Sp(4,p)

In this section Γ = Sp(4,p), a group of type C_2 (or B_2). We shall indicate how the ideas of §12 can be carried over to this case, as a preliminary to attempting a general formulation in §14.

13.1 Deformations of linkage classes

As in §12, we begin with a formal recipe which yields generic character degrees. Fix $\tau \in W$ and choose any λ in the top alcove of X_p. Deform the linkage class of λ by replacing $\lambda(\sigma)$ by $\lambda(\sigma) + \sigma\tau\sigma^{-1}(\delta_\sigma)$. Add to this list of 8 weights all the dominant weights below and linked to each of them (in general these weights all lie in X_p), for a grand total of 20. Since X_p meets 4 alcoves, each of the following configurations of weights will occur twice in the τ-deformation:

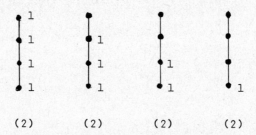

As in (12.1), the diagram indicates that the weights lie in certain alcoves and are linked as well as related by the partial order.

THEOREM. Fix $\tau \in W$. Then $\Sigma\, p\text{-dim}(\mu)$, sum over the τ-deformation associated with any big weight λ, equals the generic degree polynomial given in the following table which is associated with the conjugacy class of τ in W:

family	representative of class of τ	generic degree	frequency
A	1	$p^4+2p^3+2p^2+2p+1$	1/8
B	σ_1	p^4-1	1/4
C	σ_2	p^4-1	1/4
D	$\sigma_1\sigma_2$	p^4-2p^2+1	1/4
E	-1	$p^4-2p^3+2p^2-2p+1$	1/8

This is stated in Humphreys [6]. To prove it directly would be feasible, but lengthy: one would have to add 12 Weyl dimensions, not just 8. As an alternative method, notice that what is being asserted (for a given τ) is a polynomial identity of known degree in the variables p, r, s (where $\lambda = r\lambda_1 + s\lambda_2$), one side of which is in fact independent of r,s. So it would suffice to check that both sides take the same value for suitably many choices of p,r,s. This was done by the author using a computer algorithm developed in the first place to experiment with several possible deformation patterns.

Although the patterns just described are purely formal, it can be confidently expected (as for SL(3,p)) that the actual reduction modulo p of the Z_i behaves concordantly.

13.2 Picturing PIM's

Guided by the method of (12.2), we begin with a weight $\lambda = (r,s)$ in the top alcove, whose PIM generically has dimension $8p^4$. The composition factors involved can be deduced from the tensor product construction, cf. (9.2); in particular, there are generically 160 of these, occurring in the 8 deformation patterns described in (13.1). The weight λ itself occurs 8 times, while 4 other weights in the top alcove each occur twice. To depict the "big" weights conveniently, we surround a vertex labelled λ with 8 small copies of the top alcove, one for each element of W, and attach big weights to the corners of the resulting square according to the deformation patterns in which they occur. For example, let $p = 11$, $\lambda = (9,4)$ (here α_1 is long, α_2 is short), to get the picture below.

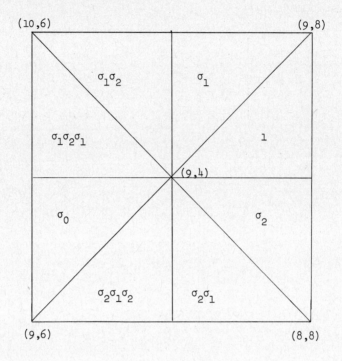

The available evidence indicates that PIM's corresponding to weights in the other 3 alcoves of X_p should be pictured as indicated below. The only vertices labelled with weights should be the "special" points (cf. Bourbaki [1, V, 3.10]), i.e., the vertices through which all possible types of reflecting hyperplanes pass.

13.3 The Brauer complex

Unless $p = 2$, there are always 2 blocks of highest defect, since $f = 2$. To construct the Brauer complex of a block (corresponding to a choice of one coset of X/X_r), begin with the top alcove of X_p and pass reflecting hyperplanes through each lattice point (this creates new vertices, which will not be labelled with weights, as noted in (13.2)). Let $W^1 = \{1,\sigma_2\sigma_1\sigma_2\}$: these are the elements of W linking

16p^4

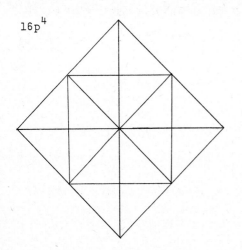

24p^4

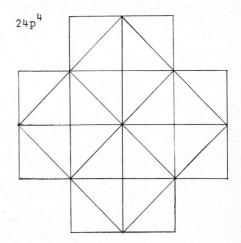

32p^4

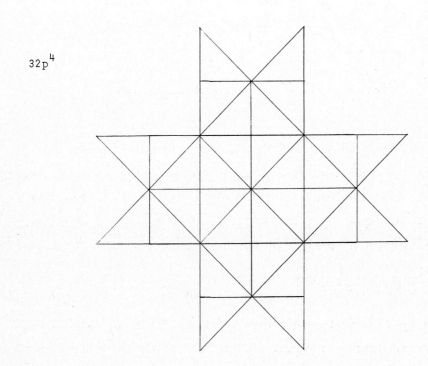

one weight in the top alcove to another. The corresponding weights δ_σ are $(0,0)$ and $(0,1)$. Given a weight λ in the top alcove, there is a unique $\sigma \in W^1$ for which $\lambda - \delta_\sigma$ lies in the chosen coset of X/X_r; replace λ by $\lambda(\sigma)$.

This recipe is like the one given in (12.3). It of course has to be checked that this leads to the same pictures of individual PIM's as given in (13.2). Notice that (for $p \neq 2$), one part of the complex is a mirror image of the other; the behavior along the "mirror" is degenerate, as will be pointed out in (13.4). The case $p = 3$ is illustrated below, although this is too small to indicate the regularities which eventually predominate; the case $p = 11$ has also been worked out in detail by the author. It is quite instructive to compare the Brauer complex with the character table of Γ in Srinivasan [2] or Springer [3]. (Cf. also the table of Cartan invariants in Humphreys [5].)

13.4 Degeneracies

When $p \neq 2$, we have already mentioned that the Brauer complex involves mirror images (the mirror running from the right angle "corner" to the "hypotenuse" wall). The alcoves of types A, B, D, E along the mirror correspond to pairs of Z_i having the same Brauer character.   This behavior shows up clearly in the character table.

At a wall, or intersection of two walls, degeneracies occur systematically. For example, along the hypotenuse wall each alcove labelled A (except the one at the corner) corresponds to a pair of Z_i of respective dimensions $p(p+1)(p^2+1)$, $(p+1)(p^2+1)$. If we had continued the complex across that wall we would have gotten an alcove of type B; the degrees just listed are equal to (respectively) half the sum and half the difference of the generic degrees of types A, B. This reveals the influence of the subgroup of order 2 in W generated by the reflection in this wall, or rather of its character table (as in (12.4)). Similarly, at the right angle corner, the character

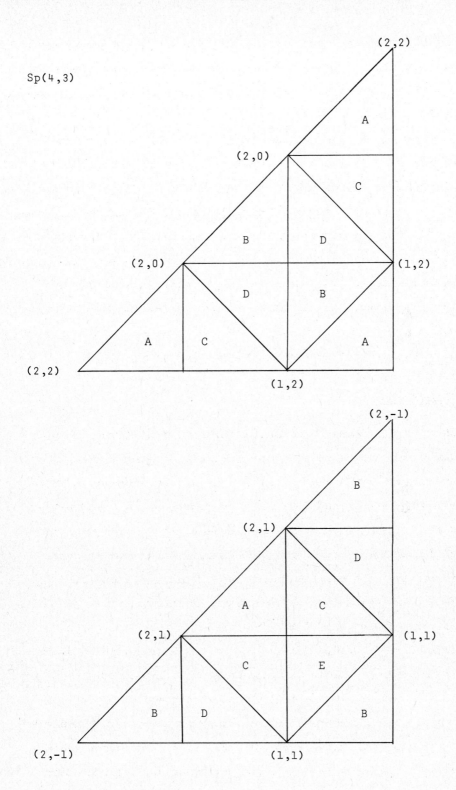

table of a Klein 4-group in W comes into play (along with the mirror-
ing phenomenon).

However, at the top corner of the complex (for the block involving
(0,0)), the character table of W does not predict the actual constituents of the representation induced from the 1-character of a Borel
subgroup (cf. (12.4)). This remains to be analyzed more fully.

There is one Z_i, denoted θ_{10} by Srinivasan, which appears to occur
in the PIM of (0,0) in a rather subtle way; its degree is $p(p-1)^2/2$,
and it can be associated with the two "discrete series" families D, E.
In the framework of Deligne, Lusztig [1], θ_{10} is obtained as a constituent of a _virtual_ character associated with the 1-character of a
suitable torus.

§14. The general case

In this section we shall attempt to describe the relationship
between ordinary and modular representations of Γ in the general case.
What we can say here is necessarily somewhat tentative, since neither
the ordinary nor the modular theory has yet been fully worked out.

14.1 Deformations of linkage classes

The nature of the ordinary representations of Γ (or, more generally, any finite group of Lie type) has been explored very success-
fully in the recent work of Lusztig [4], Deligne and Lusztig [1]. In
particular, they construct virtual representations R_T^θ parametrized by
the maximal tori T of G defined over the finite field in question and
by the linear characters θ of such tori, so that all irreducible
representations are (in principle) obtainable as constituents of these,
cf. Deligne, Lusztig [1, 7.7]. (In our notation, T has been used to
denote just the split maximal torus of G; we temporarily adopt the

broader meaning.) The formal degree of R_T^θ is given by $\pm |\Gamma|/p^m |T(\mathbb{F}_p)|$,
a polynomial in p with lead term p^m, and when θ is in "general position"
$\pm R_T^\theta$ is actually an irreducible representation. Here $|T(\mathbb{F}_p)|$ means the
order of the group of rational points of T over $\mathbb{F}_p$. The decomposition
of R_T^θ has not yet been obtained in all cases; but very recently Lusztig
has been able to treat the case of a "Coxeter torus".

The $\mathbb{F}_p$-conjugacy classes of maximal tori of G defined over $\mathbb{F}_p$ are
in natural 1-1 correspondence with the conjugacy classes of W, so the
virtual representations R_T^θ fall into families parametrized by the
classes of W, each family having a certain generic degree (indicated
above). Moreover, the family corresponding to the class of $\tau \in W$
accounts for roughly $1/|C_W(\tau)|$ of all irreducible representations of Γ.
For example, the family corresponding to $\tau = 1$ consists of the representations induced from irreducible (1-dimensional) representations of a
Borel subgroup. All of this is well illustrated by the cases treated
in §§11, 12, 13.

We have been pursuing an analogy between the $\underline{u}$-modules Z_λ and the
$K\Gamma$-modules $\bar{Z}_i$ (at least, those having the generic degrees: more
precisely, we should consider the reduction modulo p of $\pm R_T^\theta$). Typically, the highest weights of composition factors of the latter should
constitute a deformation of the linkage class of some λ. The connection between λ and θ should be as indicated earlier in special cases,
e.g., when λ lies in the top alcove and $Q_\lambda = R_\lambda$ is constructed in
$M_\mu \otimes St$, the weight μ defines characters θ of the various tori T and the
corresponding $\pm R_T^\theta$ should occur in R_λ. We note that the dimensions
involved here are in general compatible, since the generic character
degrees (one for each element of W) add up to $|W| p^m$, cf. identity (5)
in Solomon [1].

Precise deformation patterns like those exhibited earlier for
types A_1, A_2, B_2 (= C_2) cannot be hoped for until one has fairly complete information about the modules M_λ. However, it is clear from the

tensor product construction (cf. (9.2) in particular) that a good
starting point is to deform the linkage class of λ by replacing $\lambda(\sigma)$ by
$\lambda(\sigma) + \sigma\tau\sigma^{-1}(\delta_\sigma)$. (Indeed, for the series of representations corresponding to $\tau = 1$, Carter and Lusztig [2] have shown that these sets of
deformed weights always do occur.) It might be hoped that all other
required highest weights will be linked to these and below them in the
partial order. But there is a complication in case nonrestricted
dominant weights lie below restricted ones, which shows up in G_2, A_3,
etc. Here one must use deformed versions of such weights, which for Γ
look like certain other weights in X_p. This complication appears to be
unavoidable, but perhaps manageable.

Recent work of Jantzen [4] on an apparently unrelated decomposition
problem suggests a precise general conjecture (modulo a knowledge of the
M_λ). In this paper he observes generic decomposition behavior for $\bar{V}_\lambda$
(as G-module) when λ is <u>not</u> restricted but is in sufficiently general
position inside a higher alcove of X^+, relative to p^2 and not p. The
number and placement of highest weights of composition factors of $\bar{V}_\lambda$ is
essentially independent of λ, and (via Verma's Conjecture IV, now
proved) the total number of such weights is the same as for a typical
Z_λ, cf. §5. Jantzen summarizes the generic decomposition behavior in
low ranks by means of diagrams, indicating the number of weights
occurring in each alcove and the partial order among them. The number
in parentheses indicates the number of times the pattern occurs. We
have already used these diagrams in (12.1) and (13.1), because they
coincide precisely with our previously discovered deformation patterns.
His diagrams for type A_3 are given below (he also treats G_2). Notice
here that not all multiplicities are equal to 1 (so we expect that the
"decomposition numbers" for $K\Gamma$ in Brauer's sense will not always be
generically equal to 1). The last diagram includes contributions from
the two nonrestricted alcoves occurring below the top alcove of X_p, in
line with our comments above.

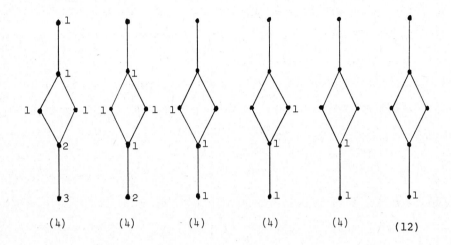

We conjecture that Jantzen's diagrams (which can be effectively computed once one knows the M_λ) also describe the generic decomposition behavior of the $\bar{Z}_i$ (or R_T^θ). This conjecture is made reasonable by the method he uses, which starts with the theorem of Hulsurkar [1] and replaces the dimension calculations with formal character calculations. (This leads to a proof of Verma's Conjecture IV; Verma himself has used essentially the same method in earlier, as yet unpublished, work.)

In spite of substantial progress, there is still no general description of the weight multiplicities of M_λ. So it may also be desirable to attempt a description of $\bar{Z}_i$ via the formal characters $\text{ch}(\lambda)$ rather than p-ch(λ). Verma [3] has suggested such a description, based on the formal character calculations just mentioned. This appears to yield the right formal degrees.

14.2   The Brauer complex

All evidence points to the conclusion that in higher ranks a "typical" $\bar{Z}_i$ will have a great many composition factors, conjecturally the sum of all d_λ (λ running over a full linkage class). In order to portray effectively the incidence relation involving the $\bar{Z}_i$ and the R_λ

(or M_λ), some composition factors should be viewed as being more significant than others. We shall emphasize the weights in the top alcove (and its closure), whose PIM's are "small" (9.2). For λ in this alcove, $d_\lambda = 1$ unless p divides f, so the linkage class of λ will contain precisely f weights belonging to the top alcove. Apart from type A_ℓ, f is of course quite small.

As suggested already for types A_1, A_2, B_2, we want to assign to each block of highest defect in $K\Gamma$ an ℓ-dimensional chamber complex. Begin with the top alcove meeting X_p (an ℓ-simplex), whose lattice points are to be the vertices of our complex. Pass through each lattice point all possible hyperplanes parallel to the walls of the alcove, thereby partitioning the alcove into small alcoves of the same type. Some new vertices may be introduced in this way (cf. B_2), but they will not be labelled with weights. The original vertices are precisely the special points of the complex (cf. Bourbaki [1, V, 3.10; VI, 2.2, 2.3]), f of which occur as vertices of each small alcove.

Now choose a coset of X/X_r corresponding to the chosen block; if more than one coset corresponds to this block, the choice of another turns out merely to re-orient the complex as we shall describe it. Let W^1 be the subgroup of W which links weights in the top alcove to other weights in the top alcove. It can be checked that $\sigma \mapsto \delta_\sigma + X_r$ sets up a bijection between W^1 and X/X_r. Given λ in the top alcove (or its closure), $\lambda - \delta_\sigma$ belongs to the chosen coset for a unique $\sigma \in W^1$; we then label the vertex λ by the σ-linked weight. This procedure replaces weights on the walls of the top alcove by others of the same sort. The weights which now appear at vertices are just those "belonging" to the chosen block, each repeated as many times as the number of blocks of highest defect.

Consider a single small alcove, with a vertex labelled λ. The other special vertices of the alcove (if any) will then be labelled with the other "big" weights which occur in some deformation

$\lambda \mapsto \lambda(\sigma) + \sigma\tau\sigma^{-1}(\delta_\sigma)$ of the linkage class of λ. In this way the small alcoves having λ as a vertex are associated with the elements τ of W. This assignment of τ to a small alcove depends on the vertex λ; but for another vertex μ we would simply replace τ by a conjugate in W. To verify all of these assertions, it may be most convenient to pass from the weight λ in the top alcove to the "dual" weight $\mu = (p-1)\delta + \sigma_o(\lambda)$ in the bottom alcove, μ being the weight involved in the tensor product construction of PIM's in (8.2). Then the big weight linked to λ by σ is "dual" to $\tau\mu + p\delta_\tau$, $\tau = \sigma_o\sigma\sigma_o$. A formulation like this has been suggested by Jantzen.

The $\bar{Z}_i$ (or, more generally, the reductions modulo p of certain of the virtual representations R_T^θ) should now be assigned to various small alcoves, depending on the various conjugacy classes of W as indicated above. To justify this assignment one would have to see how different PIM's overlap, which would in turn require detailed knowledge of PIM's (not just those belonging to big weights). What is clear at this point from a comparison of the Brauer characters in (9.2) and the character formulas in Deligne, Lusztig [1], is that a "small" PIM $Q_\lambda = R_\lambda$ will involve the various R_T^θ if μ is dual to λ as above and determines the character θ of T. It seems likely that to each small alcove in the complex can be unambiguously assigned an R_T^θ corresponding to one of the p^ℓ "geometric conjugacy" classes of pairs (T,θ) described in §5 of their paper.

14.3 Degeneracies

In case there is more than one block of highest defect, all vertices of the Brauer complex of a given block will be repeated, cf. type A_2 when $p \equiv 1 \pmod 3$, or type B_2 when p is odd. Then it seems likely that one or more alcoves will have repeated vertices, corresponding to the presence there of several distinct $\bar{Z}_i$'s whose composition factors (and hence Brauer characters) are the same. This kind of ramification is perhaps describable in terms of an action of the center of Γ.

The most interesting degeneracies in the Brauer complex of a block occur at the walls. Here some PIM's become truncated (cf. the regularity conjecture of (10.3)), while others involve nonstandard mixtures of the $\bar{Z}_i$. To understand what happens in these cases, it may be desirable to think in terms of continuing the complex beyond its walls, labelling the small alcoves with virtual representations R_T^θ according to the characters θ involved. For example, only one small alcove with vertex $(p-1)\delta$ (dual to the weight 0) actually occurs inside the complex, but the representations R_T^1 are defined for all types of tori (not just the split torus); here 1 is the character of T arising from the weight 0. When a small alcove is bounded by only one of the outside walls of the complex, Jantzen has suggested a general analysis along the lines of the special cases treated in (12.4) and (13.4). Where two or more walls intersect, the analysis will of course become more complicated. The corresponding subgroup of W (generated by the reflections in these walls) ought to play a leading role in such an analysis.

14.4 A duality phenomenon

We conclude this section by pointing out an interesting sort of duality between PIM's and Brauer liftings. Formula (5) in Solomon [1] can be interpreted as providing, for each compound of the reflection character of W, a formal identity involving the generic character degrees for Γ (or Γ_n); one has to specialize t to p (or p^n). For example, the 1-character of W corresponds to the identity: $|W|p^m =$ sum of generic degrees associated with the various conjugacy classes of maximal tori over $\mathbb{F}_p$, each taken as often as the number of elements in the corresponding conjugacy class of W. This identity arises more concretely from a consideration of the $\bar{Z}_i$ which typically occur in a PIM of dimension $|W|p^m$.

The alternating character ε of W corresponds to the identity: |W| = sum of generic degrees as above, weighted by ±1 according to the class of W involved. (For type A_1, this amounts to: 2 = (p+1)-(p-1).) Experimental evidence in low ranks shows that this identity also has a concrete interpretation: adding up the Brauer characters of the $\bar{Z}_i$ which occur in a typical small PIM $Q_\lambda \subset M_\mu \otimes St$ (9.2), with the indicated coefficients ±1, leads to the virtual Brauer character s'(μ) obtained from the orbit sum as in (10.4). This procedure works even when the W-orbit of μ is smaller than |W|; here fewer $\bar{Z}_i$ will occur, of course.

The two identities just discussed arise (in degenerate form) in the work of Lusztig [2, 5.6] on the Brauer lifting for GL(V) over $\mathbb{F}_q$, where V is of any finite dimension. As mentioned in (10.3), he shows that for p ≠ 2, V⊗St is a PIM for GL(V). The ordinary characters which are constituents of this principal indecomposable character are just the terms in his alternating sum expression for the Brauer lifting of the representation V. This adapts to the special linear group as well, where the highest weight of V is just the first fundamental weight and where the formal character is just the orbit sum under W of this weight. The formal duality involved in his calculations is shown in the two identities at the top of p. 88 of his paper. He and Verma worked out further examples of this duality as well.

It seems likely that for any PIM, the alternating sum of Brauer characters of the $\bar{Z}_i$ occurring will be a sum of one or more virtual Brauer characters s'(μ) in a natural way. Perhaps this is provable directly for "small" PIM's by use of the character formulas in Deligne, Lusztig [1], cf. their Proposition 7.11. We can give a heuristic argument based on formal Weyl characters (as in the proof of Theorem 9.2), as follows. Construct $Q_\lambda = R_\lambda$ in $M_\mu \otimes St$ as in (9.2), so that $\mu = \lambda^\circ + \delta$ in Λ. Assuming that deformation patterns of the sort conjectured in (14.1) describe the composition factors of $\bar{Z}_i$, it should be true that the contributions to the alternating sum of $\bar{Z}_i$'s made by the M_ν for ν in alcoves other than the lowest one will cancel each other. This is due to the fact that δ_σ is irregular (has nontrivial stabilizer in W) unless $\sigma = \sigma_o$, cf. Table 2 in (12.1). For ν in the lowest alcove, we have $M_\nu = \bar{V}_\nu$ (4.1). Therefore the alternating sum calculation reduces to:

$$\sum_\tau \epsilon(\tau) \, ch(\lambda^o + \sigma_o \tau \sigma_o(\delta))$$

$$= \sum_\tau \epsilon(\tau) \, ch(\lambda^o + \tau(\delta))$$

$$= \sum_\tau \epsilon(\tau) \, ch(\mu - \delta + \tau\delta)$$

$$= \sum_{\sigma,\tau} \epsilon(\sigma\tau) e(\sigma(\mu + \tau\delta))/q$$

$$= \sum_{\sigma,\tau} \epsilon(\sigma\tau) e(\sigma\mu) e(\sigma\tau(\delta))/q$$

$$= \sum_\sigma e(\sigma\mu) \sum_\tau \epsilon(\sigma\tau) e(\sigma\tau(\delta))/q$$

$$= \sum_\sigma e(\sigma\mu).$$

Here we have assumed that μ has W-orbit of full size; but the degenerate cases should go through similarly.

Since the Brauer characters of the M_ν are built up from the s'(μ) (cf. Wong [1]), the conjectural method just described would lead to a rather explicit Brauer lifting, in the spirit of Lusztig [2]. Indeed, it appears (from the cases studied so far) that if the actual characters, not just Brauer characters, of the Z_i are used in the alternating sums, the resulting virtual characters will behave in the typical manner of Brauer liftings, i.e., the value of such a virtual character on an element x of Γ will equal the value on the semisimple part of x.

IV. TWISTED GROUPS

Here we shall adapt some of the results of parts I, II, III to the twisted groups of types A, D, E_6 which arise by combining a graph auto- morphism with a field automorphism. For example, start with a field of q^2 elements $(q = p^n)$, let ϕ be its qth power automorphism, and combine ϕ with a graph automorphism of order 2 to get an endomorphism σ of G having order 2. When $G = SL(\ell+1,K)$, the resulting fixed point set G_σ is the special unitary group $SU(\ell+1,q^2) \subset SL(\ell+1,q^2)$. (Cf. Steinberg [4, §11; 5, §11].) We denote this twisted subgroup of Γ_{2n} (or Γ_{3n}, as the case may be) by Γ_n'; in particular, write just Γ' in place of Γ_1'.

§15. Ordinary and modular representations

15.1 Irreducible modules

Steinberg [3; 4, §13] has determined the irreducible modular representations of Γ'.

THEOREM. Let $\lambda \in X_p$. Then the G-module M_λ remains irreducible on restriction to Γ', and the resulting set of p^ℓ modules exhausts (up to isomorphism) the irreducible $K\Gamma'$-modules.

There is also a twisted tensor product theorem for Γ_n', extending (2.1).

15.2 Projective modules

PROPOSITION. Let $\lambda = (p-1)\delta$. Then $M_\lambda = St$ is a projective $K\Gamma'$-module.

Proof. This is proved in the same way as Proposition 10.1, using the fact that $\mathbb{C}\Gamma'$ has an irreducible representation of degree p^m (cf. Steinberg [4, 15.5]).

Similarly, $M_{(p^n-1)\delta}$ is a projective $K\Gamma'_n$-module.

Denote by R'_λ ($\lambda \in X_p$) the PIM of Γ' having M_λ as its top and bottom composition factor. It is well known that $|\Gamma'|$ is a polynomial in p divisible by p^m (but no higher power of p), so p^m divides the dimensions of all R'_λ (R.4). According to Lemma 8.1 and the above proposition, $M_\mu \otimes \mathrm{St}$ is a projective $K\Gamma'$-module ($\mu \in X_p$).

THEOREM. Let $\lambda \in X_p$. The G-module Q_λ of (8.2) is a projective $K\Gamma'$-module, having R'_λ as a direct summand.

Proof. The proof is like that of Theorem 10.2. Q_λ is a G-summand of the projective $K\Gamma'$-module $M_\mu \otimes \mathrm{St}$ (for suitable μ), and has a G-submodule isomorphic to M_λ, so it is a direct sum of PIM's for $K\Gamma'$, one of these being R'_λ.

As in (10.3), we can conjecture that whenever λ is regular, $R'_\lambda = Q_\lambda$ (as $K\Gamma'$-modules). Since the dimensions of $K\Gamma$, $K\Gamma'$ are polynomials in p having highest term $p^{\ell+2m}$, and since their irreducible modules coincide, we expect that $Q_\lambda = R_\lambda = R'_\lambda$ most (but not all) of the time.

15.3 An example

The example $\Gamma' = SU(3,25)$ has been worked out by ad hoc methods. Here $|\Gamma| = p^3(p^2-1)(p^3-1)$, while $|\Gamma'| = p^3(p^2-1)(p^3+1)$.

In order to list the composition factors of Q_λ as a $K\Gamma'$-module, we have to consider how $M_\nu^{(p)}$ looks when restricted to Γ'. Raising matrix entries to the pth power corresponds to applying the (inverse of the) graph automorphism in question. It is well known that the graph automorphism coincides with $-\sigma_o$ <u>except</u> in type D_ℓ (ℓ even), where $-\sigma_o = 1 \in W$. Thanks to (2.4), we conclude that raising matrix entries to the pth power is exactly the same as passing to the contragredient module $M_\nu^* \cong M_{-\sigma_o \nu}$ (except for type D_ℓ, ℓ even). Writing $\nu = r\lambda_1 + s\lambda_2 = (r,s)$, we have $-\sigma_o \nu = (s,r)$. Now it is easy to adapt the composition factors of the small PIM's Q_λ, as listed in (9.3), to $K\Gamma'$. The required symmetry of the Cartan matrix of $K\Gamma'$ (R.3), along with considerations of dimension, can now be used to determine the Cartan invariants for the three blocks of $K\Gamma'$ of highest defect, as shown in tables below. The third block may be obtained from the second by interchange of coordinates. Each block has determinant 5^4 (as predicted by Brauer's theory).

	00	33	11	22	03	30	41	14
00	12	4	10	9	2	2	1	1
33	4	6	9	4	4	4	2	2
11	10	9	26	15	4	4	5	5
22	9	4	15	11	1	1	2	2
03	2	4	4	1	5	4	2	1
30	2	4	4	1	4	5	1	2
41	1	2	5	2	2	1	3	1
14	1	2	5	2	1	2	1	3

	20	31	01	23	43	04	12	42
20	16	6	10	9	1	4	6	5
31	6	7	9	4	2	2	5	2
01	10	9	22	13	1	0	8	2
23	9	4	13	10	0	0	5	2
34	1	2	1	0	3	2	2	0
04	4	2	0	0	2	4	1	1
12	6	5	8	5	2	1	6	2
42	5	2	2	2	0	1	2	3

From the displayed Cartan invariants we can deduce the way in which the Q_λ decompose, the cases in which $Q_{rs} \neq R_{rs}$ being as follows:

$$Q_{00} = R'_{00} \oplus R'_{40} \oplus R'_{04} \oplus R'_{44},$$
$$Q_{30} = R'_{30} \oplus R'_{44} \oplus R'_{44} \text{ (similarly for } Q_{03})$$
$$Q_{20} = R'_{20} \oplus R'_{24} \text{ (similarly for } Q_{02}),$$
$$Q_{10} = R'_{10} \oplus R'_{14} \text{ (similarly for } Q_{01}).$$

Notice that $Q_{40} = R'_{40}$ in spite of the nonregularity of $(4,0)$.

Since $|\Gamma'| - |\Gamma| = 2p^3(p^2-1)$, the general pattern which this suggests for arbitrary p is compatible with the dimension count in (R.2):

$$Q_{00} = R'_{00} \oplus R'_{p-1,0} \oplus R'_{0,p-1} \oplus St,$$
$$Q_{p-2,0} = R'_{p-2,0} \oplus St \oplus St \text{ (similarly for } Q_{0,p-2}),$$
$$Q_{r,0} = R'_{r,0} \oplus R'_{r,p-1} \text{ if } 0 < r < p-2 \text{ (similarly for } Q_{0,s}).$$

We remark that Zaslawsky [1] has also computed the Cartan 5-invariants of PSU(3,25) (among other cases).

15.4 Ordinary representations

Consider next the ordinary representations of Γ'. Since the modular theory of Γ' is quite close to that of Γ, we can expect a priori a similar closeness here. The main difference is that $M_\mu \otimes M_\nu^{(p)}$ looks like $M_\mu \otimes M_\nu$ when restricted to Γ but like $M_\mu \otimes M_\nu^*$ when restricted to Γ' (unless Γ' comes from a group of type D_ℓ, ℓ even, in which case the situation seems to be less straightforward).

Take again the example $\Gamma' = SU(3,p^2)$, $\Gamma = SL(3,p)$. In place of the deformation pattern $\lambda(\sigma) \mapsto \lambda(\sigma) + \sigma\tau\sigma^{-1}(\delta_\sigma)$ described in (12.1), we are led to a twisted version: $\lambda(\sigma) \mapsto \lambda(\sigma) - \sigma\tau\sigma^{-1}(\delta_\sigma)$. It is not hard to check that the resulting lists of 9 p-dimensions add up formally to the known "generic" character degrees for Γ': p^3-2p^2+2p-1, p^3+1,

p^3+p^2-p-1 (cf. Simpson, Frame [1]). These are obtained from the generic character degrees for Γ by changing the sign of each even power of p. Arguments like those in §12 can be used to show that these formal calculations correspond to the actual reduction modulo p of the irreducible 𝐶Γ'-modules.

The irreducible characters of 𝐶Γ' (or 𝐶Γ'$_n$) are not yet known completely, except for types $^2A_2$, $^2A_3$ (cf. Ennola [2], Simpson, Frame [1], Nozawa [1]). However, Ennola has conjectured that the character table of a unitary group should be obtainable from that of the corresponding general linear group by "changing q to -q" (this is not a precise formulation, of course), and presumably this should adapt to SU, SL. It seems highly plausible that a parallel phenomenon should exist for other twisted groups, although for type D_ℓ (ℓ even) it may be harder to formulate. Much progress has recently been made in the character theory by Deligne, Lusztig [1], cf. Srinivasan [6].

15.5 Groups of Suzuki and Ree

The twisted groups of Ree and Suzuki have not yet been examined in the modular context outlined above, but Steinberg [3] has described their irreducible modular representations in terms of those of the corresponding Chevalley group. It seems quite likely that their PIM's and then their ordinary representations can be dealt with in a parallel way.

APPENDIX R: REPRESENTATION THEORY

Here we recall a few aspects of the representation theory of
associative algebras over K, with reference to CR = Curtis, Reiner [1]
(see also Dornhoff [1, Part B], Chevalley [1]).   Let A be a finite
dimensional associative algebra over K.   By A-module we mean a left
A-module, finite dimensional over K.

R.1 Principal indecomposable modules

The principal indecomposable modules (PIM's) of A are the indecomposable left ideals of A, or equivalently, the indecomposable projec-
tive A-modules (CR, 56.6).   Let $\{U_\lambda \mid \lambda \in \Lambda\}$ be a complete collection of
nonisomorphic PIM's.   Then each U_λ has a unique maximal submodule,
hence a unique top composition factor M_λ.   The M_λ are pairwise nonisomorphic, and every irreducible A-module is isomorphic to one of
them (CR, 54.14).   Two PIM's U_λ, U_μ are said to be linked if there is
a finite chain of PIM's from U_λ to U_μ in which each adjacent pair
shares a composition factor.   The sum of all PIM's in A belonging to a
class of this equivalence relation is an indecomposable 2-sided ideal
of A, called a block, and A is the direct sum of its blocks (CR, §55).

R.2 Frobenius algebras

A is a Frobenius algebra if the A-module A and the dual of the
right A-module A (this dual viewed as a left A-module) are isomorphic
(CR, §61).   For example, take A to be the group algebra KΓ of a finite
group Γ (CR, 62.1) or to be the restricted universal enveloping
algebra of a Lie p-algebra (Berkson [1]).   If A is Frobenius, projective A-modules coincide with injective A-modules (CR, 62.11).   More-
over, if M is any A-module, dim $\mathrm{Hom}_A(U_\lambda, M)$ = number of times M_λ occurs

as a composition factor of M; while if U is a projective A-module,
dim $\text{Hom}_A(U, M_\lambda)$ = number of times U_λ occurs as a direct summand of U
(CR, 54.16, 61.13, 83.3). In particular, dim A = $\sum_{\lambda \in \Lambda} (\text{dim } M_\lambda)(\text{dim } U_\lambda)$.

R.3 Symmetric algebras

A is called symmetric if it admits a nondegenerate symmetric
associative bilinear form (then A is Frobenius), e.g., A = KΓ (CR, §66)
or A = restricted universal enveloping algebra of the Lie algebra of a
Chevalley group over K (Humphreys [1, p. 71]). Write $U_\lambda \underset{A}{\leftrightarrow} \sum c_{\lambda\mu} M_\mu$ to
indicate the composition factors of U_λ (the numbers $c_{\lambda\mu}$ are called the
Cartan invariants of A). If A is symmetric, the Cartan matrix
C = $(c_{\lambda\mu})$ is a symmetric matrix (CR, §66, 83.10), cf. Humphreys [1,
p. 71]. Moreover, each U_λ has a unique irreducible submodule, isomorphic to M_λ (CR, Ex. 83.1, which adapts to any symmetric algebra).

R.4 Group algebras

Let A = KΓ be a group algebra. If p^m is the exact power of
p = char K which divides $|\Gamma|$, then p^m divides dim R_λ for all $\lambda \in \Lambda$
(CR, 84.15). Let $Z_1, \ldots, Z_s$ (s = number of classes of Γ) be a full set
of irreducible $\mathbb{C}\Gamma$-modules; these possess reductions modulo p, say $\bar{Z}_i$,
whose composition factors do not depend on how the reduction is done
(CR, 82.1). Write $\bar{Z}_i \underset{K\Gamma}{\leftrightarrow} \sum_{\lambda \in \Lambda} d_{i\lambda} M_\lambda$, and call D = $(d_{i\lambda})$ the decomposition
matrix. Then C = $^tD\,D$, where C is the Cartan matrix (CR, 83.9). Thus
the composition factors of U_λ are those of various $\bar{Z}_i$, with multiplicity determined by the number of times M_λ occurs as a composition
factor of $\bar{Z}_i$. In particular, all composition factors of $\bar{Z}_i$ lie in a
single block of KΓ.

APPENDIX S: THE STEINBERG REPRESENTATION

If a finite group has an irreducible complex character whose degree is divisible by the order of a p-sylow subgroup, then this character vanishes off the p-regular classes and remains irreducible upon reduction modulo p (where it is also a principal indecomposable character). This is a basic result of Brauer, the case of a block of "defect 0", cf. Curtis, Reiner [1, 86.3] or Dornhoff [1, Part B, §62].

It was already appreciated by Schur [1] that Γ_n = SL(2,p^n) has an irreducible complex character of degree p^n, the highest power of p dividing the group order. This character, whose values on p-regular (= semisimple) classes are integers ±1, p^n, can be obtained as a constituent of the character of Γ_n induced from the 1-character of the upper triangular group.

In his thesis, Steinberg [1] found an analogous character χ for some other finite linear groups. In Steinberg [2,3,4] this is generalized to the case of an arbitrary finite Chevalley group and then to the twisted groups as well. For a Chevalley group Γ_n over the field of $q = p^n$ elements, χ has degree q^m (m = number of positive roots). It occurs once as a constituent of the character induced from the 1-character of a Borel subgroup, but can be obtained more concretely by constructing a minimal left ideal in the complex group algebra. A generator for this left ideal is $\Sigma\epsilon(w)wb$, where b runs over the standard Borel subgroup and w runs over a set of representatives of the Weyl group (ϵ being the alternating character of W).

Brauer's theorem insures that the Steinberg character χ vanishes off p-regular classes. Using his explicit construction, Steinberg showed that the character value at a semisimple element x is $\pm q^{d(x)}$, where $q^{d(x)}$ is the order of a p-sylow subgroup of the centralizer of x and the sign ± is not specified (cf. Steinberg [4, 15.5]). Later,

Srinivasan [3] showed (except for certain p) that the ±1 is just ε(w), where the class of w in W corresponds to the conjugacy class of maximal tori of the centralizer of x defined over the finite field in question. Her restriction on p apparently can now be removed because of the work of Deligne, Lusztig [1].

Brauer's theorem also insures that the Steinberg representation (say of Γ_n) remains irreducible modulo p, where it must coincide with some M_λ ($\lambda \in X_q$). Comparison of dimensions shows that $\lambda = (q-1)\delta$, cf. (2.2); we write St_n for M_λ in this case. Steinberg's tensor product theorem (2.1) further shows that St_n is a twisted tensor product built up from copies of St, a fact which enters (perhaps avoidably) into his character calculation at one point. In addition, St_n is a PIM for Γ_n; St also turns out to be a PIM for u, cf. (5.5).

Curtis [3] and Feit independently observed that the Steinberg character (of Γ_n or twisted analogue) could be expressed as the alternating sum of the characters induced from the 1-characters of standard parabolic subgroups. With the aid of Tits, Solomon [3] then gave a homological interpretation of this formula, by showing that the Steinberg representation is realizable over $\mathbb{Z}$ as the top (reduced) homology of the Tits building, the latter being homotopy equivalent to a bouquet of ℓ-1 spheres. It is noteworthy that all of this reduces to the alternating character of W when q is specialized to 1. It is also noteworthy that the analogous construction for p-adic groups leads to the "special" representation (Borel-Serre, Garland).

APPENDIX T: TENSORING WITH A PROJECTIVE MODULE

It has long been recognized that if A is the group algebra of a finite group and P is a projective A-module, then $P \otimes M$ is also projective (M being an arbitrary A-module), cf. Curtis, Reiner [1, p. 426, ex. 2], Alperin [1]. The fact that the tensor product is again an A-module depends, of course, on the Hopf algebra structure of A arising from group multiplication.

For the restricted universal enveloping algebra of a Lie p-algebra the same principle holds, cf. Pareigis [1, Lemma 2.5] or Humphreys [4]. As in the case of a group algebra, the tensor products make sense because of a Hopf algebra structure (arising from the Lie algebra).

M. Sweedler has communicated to the author a general argument applicable to any Hopf algebra H (with antipode) over a commutative ring, no special finiteness assumptions being required. If M is any left H-module, the tensor product $H \otimes M$ (over the base ring) has two H-module structures, one induced by left multiplication on H and the other induced by the coproduct on H; in the latter case, $h \cdot (h' \otimes m)$ = $\Sigma\, h_{(1)} h' \otimes h_{(2)} \cdot m$, if $h \mapsto \Sigma h_{(1)} \otimes h_{(2)}$. The key observation is that $h \otimes m \mapsto \Sigma h_{(1)} \otimes h_{(2)} \cdot m$ defines an H-module isomorphism between the two structures, whose inverse is given explicitly by $h \otimes m \mapsto \Sigma h_{(1)} \otimes s(h_{(2)}) \cdot m$, s being the antipode. (In Humphreys [4], a dimension argument is used in place of this explicit description of the inverse map.) Now if M is a projective module over the base ring (e.g., a vector space over K), the first H-module structure on $H \otimes M$ is clearly that of a projective H-module, thought of as an "induced" module. The above isomorphism implies that the more complicated H-module structure on $H \otimes M$ is also that of a projective H-module. To complete the argument, one just has to pass from H to a direct sum of copies of H and then to a direct summand.

Conclusion: Tensoring an arbitrary H-module with a projective H-module

yields a projective H - module.

APPENDIX U: THE UNIVERSAL ENVELOPING ALGEBRA

Here we outline briefly some constructions related to the universal enveloping algebra U of $\underline{g}_\mathbb{C}$. Complete details will appear elsewhere.

Kostant's $\mathbb{Z}$-form $U_\mathbb{Z}$ is defined to be the subring of U generated by all $X_\alpha^t/t!$ and all $Y_\alpha^t/t!$. This turns out to have a nice $\mathbb{Z}$-basis analogous to the usual Poincaré-Birkhoff-Witt basis of U. For details, cf. Kostant [1], Steinberg [5], Humphreys [3]. $U_\mathbb{Z}$ has a natural structure of Hopf algebra with antipode, which carries over to the "hyperalgebra" $U_K = U_\mathbb{Z} \otimes K$. This algebra is used in the work of Carter, Lusztig [1], Jantzen.

Kostant originally used $U_\mathbb{Z}$ to define a $\mathbb{Z}$-form of G via an intrinsically defined affine algebra $\mathbb{Z}[G]$. Passing to K, one can apparently show that U_K acts naturally on $K[G]$, leaving stable the same subspaces that G leaves stable (acting by right translations). It is well known that any G-module can be embedded in a finite direct sum of copies of the G-module $K[G]$, so it will follow that G-modules are essentially the same thing as U_K-modules. Moreover, G acts naturally on U_K itself as a group of algebra automorphisms, compatible with the adjoint representation on $\underline{g}$.

For each power $q = p^n$, we can define a subalgebra $\underline{u}_n$ of U_K having dimension $q^{\dim G}$, as suggested by Verma: $\underline{u}_n$ is to be generated by all $X_\alpha^t/t! \otimes 1$ and all $Y_\alpha^t/t! \otimes 1$ for which $t < q$. For example, $\underline{u}_1$ is the subalgebra of U_K generated by $\underline{g}$, and is identifiable with the restricted universal enveloping algebra $\underline{u}$. The algebra $\underline{u}_n$ inherits from U_K the structure of Hopf algebra with antipode. Moreover, G acts naturally on $\underline{u}_n$.

Now one wants to introduce for $\underline{u}_n$ various modules analogous to those already studied for $\underline{u}$, to be indexed by homomorphisms of the "diagonal" subalgebra, which are essentially the weights λ in X_q. Begin with the largest possible module generated by a maximal vector of weight λ, denoted $Z_{\lambda,n}$ (of dimension q^m). Its unique irreducible homomorphic image is identifiable with the G-module M_λ (which one could denote $M_{\lambda,n}$ to be consistent); these M_λ exhaust the isomorphism classes of irreducible $\underline{u}_n$-modules. In turn, the PIM's of $\underline{u}_n$ are the G-modules $Q_{\lambda,n}$ constructed in (10.2). The entire argument for $\underline{u}$ in §8 can apparently be imitated for $\underline{u}_n$: St_n is a projective $\underline{u}_n$-module, so tensoring it with any other module yields a projective module (cf. Appendix T).

BIBLIOGRAPHY

J. L. Alperin

    1. Projective modules and tensor products, preprint.

J. L. Alperin, G. J. Janusz

    1. Resolutions and periodicity, Proc. Amer. Math. Soc. 37(1973), 403-406.

J. S. Andrade

    1. La représentation de Weil de $GSp(4,\mathbb{F}_q)$ et les représentations de $Sp(4,\mathbb{F}_q)$, C.R. Acad. Sci. Paris 278 (1974), 321-324.

J. W. Ballard

    1. On the principal indecomposable modules of finite Chevalley groups, Ph.D. thesis, U. Wisconsin 1974.

    2. Some generalized characters of finite Chevalley groups, preprint.

M. Benard

    1. On the Schur indices of characters of the exceptional Weyl groups, Ann. of Math. 94(1971), 89-107.

C. T. Benson, C. W. Curtis

    1. On the degrees and rationality of certain characters of finite Chevalley groups, Trans. Amer. Math. Soc. 165(1972), 251-273.

C. T. Benson, L. C. Grove

    1. The symmetric square and generic algebras of Chevalley groups, Comm. Algebra 1 (1974), 503-516.

C. T. Benson, L. C. Grove, D. B. Surowski

    1. Semilinear automorphisms and dimension functions for certain characters of finite Chevalley groups, Math. Z. 144 (1975), 149-159.

A. Berkson

    1. The u-algebra of a restricted Lie algebra is Frobenius, Proc. Amer. Math. Soc. 15(1964), 14-15.

I. N. Bernstein, I. M. Gel'fand, S. I. Gel'fand

    1. Structure of representations generated by vectors of highest weight, Funkcional. Anal. i Prilozen. 5, no. 1 (1971), 1-9 = Functional Anal. Appl. 5(1971), 1-8.

    2. Differential operators on the principal affine space and investigation of g-modules, in: Lie Groups and Their Representations (Proc. Summer School on Group Representations of the Bolyai János Math. Soc., Budapest, 1971), pp. 21-64, New York: Halsted 1975.

    3. A new model of the representations of finite semisimple algebraic groups, Uspehi Mat. Nauk 29, no. 3 (1974), 185-186.

A. Borel

    1. Properties and linear representations of Chevalley groups, in: Seminar on Algebraic Groups and Related Finite Groups, Lect. Notes in Math. 131, Springer 1970.

    2. Linear representations of semi-simple algebraic groups, in: Proc. Symp. Pure Math. 29, pp. 421-440, Providence, R.I.: American Math. Soc. 1975.

N. Bourbaki

    1. Groupes et algèbres de Lie, Chap. IV - VI, Hermann, Paris 1968

B. Braden

    1. Restricted representations of classical Lie algebras of type A_2 and B_2, Bull. Amer. Math. Soc. 73(1967), 482-486.

R. Brauer

    1. Sur la multiplication des caractéristiques des groupes continus et semi-simples, C.R. Acad. Sci. Paris 204(1937), 1784-1786.

R. Brauer, C. Nesbitt

    1. On the modular characters of groups, Ann. of Math. 42(1941), 556-590.

N. Burgoyne

1. Modular representations of some finite groups, in: Proc. Symp.
   Pure Math. 21, pp. 13-18, Amer. Math. Soc. 1971.

R. W. Carter

1. Simple Groups of Lie Type, New York: Interscience 1972.

2. Conjugacy classes in the Weyl group, Comp. Math. 25(1972), 1-59.

R. W. Carter, E. Cline

1. In preparation.

R. W. Carter, G. Lusztig

1. On the modular representations of the general linear and
   symmetric groups, Math. Z. 136 (1974), 193-242.

2. Modular representations of finite groups of Lie type, to
   appear.

B. Chang

1. The conjugate classes of Chevalley groups of type (G_2), J.
   Algebra 9 (1968), 190-211.

B. Chang, R. Ree

1. The characters of $G_2(q)$, Symposia Mathematica XIII (INDAM),
   pp. 395-413, London-New York: Academic Press 1974.

C. Chevalley

1. Théorie des blocs, Sém. Bourbaki 1972/73, Exp. 419, Lect.
   Notes in Math. 383, Springer 1974.

C. W. Curtis

1. Representations of Lie algebras of classical type with applications to linear groups, J. Math. Mech. 9(1960), 307-326.

2. Irreducible representations of finite groups of Lie type,
   J. Reine Angew. Math. 219 (1965), 180-199.

3. The Steinberg character of a finite group with a (B,N) pair,
   J. Algebra 4 (1966), 433-441.

4. Modular representations of finite groups with split (B,N)-pairs,
   in: Seminar on Algebraic Groups and Related Finite Groups, Lect.
   Notes in Math. 131, Springer 1970.

C. W. Curtis

    5. On the values of certain irreducible characters of Chevalley
       groups, Symposia Mathematica XIII (INDAM), pp. 343-355, London-
       New York: Academic Press 1974.

    6. Corrections and additions to "On the degrees and rationality of
       certain characters of finite Chevalley groups", Trans. Amer.
       Math. Soc. 202 (1975), 405-406.

    7. Reduction theorems for characters of finite groups of Lie type,
       J. Math. Soc. Japan 27(1975), 666-688.

C. W. Curtis, N. Iwahori, R. Kilmoyer

    1. Hecke algebras and characters of parabolic type of finite
       groups with (B,N)-pairs, Publ. Math. I.H.E.S. 40(1971), 81-116.

C. W. Curtis, W. M. Kantor, G. M. Seitz

    1. The 2-transitive permutation representations of Chevalley
       groups

C. W. Curtis, I. Reiner

    1. Representation Theory of Finite Groups and Associative Algebras,
       New York: Interscience 1962.

S. W. Dagger

    1. A class of irreducible characters for certain classical groups,
       J. London Math. Soc. 2 (1970), 513-520.

    2. On the blocks of the Chevalley groups, J. London Math. Soc. 3
       (1971), 21-29.

P. Deligne, G. Lusztig

    1. Representations of reductive groups over finite fields

J. Dixmier

    1. Certaines représentations infinies des algèbres de Lie semi-
       simples, Sém. Bourbaki 1972/73, Exp. 425, Lect. Notes in Math.
       383, Springer 1974.

    2. Algèbres enveloppantes, Paris: Gauthier-Villars 1974.

L. Dornhoff

    1. Group Representation Theory, New York: M. Dekker, Part A, 1971;
       Part B, 1972.

V. Ennola

    1. Conjugacy classes of the finite unitary groups, Ann. Acad.
       Scien. Fenn. 313 (1962), 313-326.

    2. Characters of finite unitary groups, Ann. Acad. Scien. Fenn.
       323 (1963), 120-155.

H. Enomoto

    1. The conjugacy classes of Chevalley groups of type (G_2) over
       finite fields of characteristic 2 or 3, J. Fac. Sci. Univ.
       Tokyo 16 (1969), 497-512.

    2. The characters of the finite symplectic group $Sp(4,q)$, $q = 2^f$,
       Osaka J. Math. 9(1972), 75-94.

J. S. Frame

    1. The characters of the Weyl group E_8, in: Computational Problems
       in Abstract Algebra (Oxford conference 1967), ed. J. Leech,
       pp. 111-130.

    2. Characters and classes of the groups $O_n(2)$, preprint.

I. M. Gel'fand, M. I. Graev

    1. Construction of irreducible representations of simple algebraic
       groups over a finite field, Dokl. Akad. Nauk SSSR 147 (1962),
       529-532 = Soviet Math. Dokl. 3(1962), 1646-1649.

S. I. Gel'fand

    1. Representations of the full linear group over a finite field,
       Mat. Sbornik 83 (1970), 15-41 = Math. USSR - Sb. 12 (1970),
       13-40.

    2. Representations of the general linear group over a finite field,
       in: Lie Groups and Their Representations (Proc. Summer School
       on Group Representations of the Bolyai János Math. Soc.,
       Budapest, 1971), pp. 119-132, New York: Halsted 1975.

J. A. Green

    1. The characters of the finite general linear groups, Trans. Amer. Math. Soc. 80 (1955), 402-447.

    2. On the Steinberg characters of finite Chevalley groups, Math. Z. 117 (1970), 272-288.

J. A. Green, G. I. Lehrer

    1. On the principal series characters of Chevalley groups and twisted types, Quart. J. Math. Oxford Ser., to appear.

J. A. Green, G. I. Lehrer, G. Lusztig

    1. On the degrees of certain group characters, Quart. J. Math. Oxford Ser., to appear.

W. J. Haboush

    1. Reductive groups are geometrically reductive, Ann. of Math. 102(1975), 67-83.

Harish-Chandra

    1. Eisenstein series over finite fields, in: Functional analysis and related fields, ed. F. E. Browder, pp. 76-88, Springer 1970.

A. Helversen-Pasotto

    1. Série discrète de $GL(3, \mathbb{F}_q)$ et sommes de Gauss, C. R. Acad. Sci. Paris 275(1972), 263-266.

P. Hoefsmit

    1. Representations of Hecke algebras of finite groups with BN-pairs of classical type, Ph.D. thesis, U. British Columbia, 1974

R. B. Howlett

    1. On the degrees of Steinberg characters of Chevalley groups, Math. Z. 135 (1974), 125-135.

S. G. Hulsurkar

    1. Proof of Verma's conjecture on Weyl's dimension polynomial, Invent. Math. 27(1974), 45-52.

J. E. Humphreys

1. Modular representations of classical Lie algebras and semi-simple groups, J. Algebra 19 (1971), 51-79.

2. Defect groups for finite groups of Lie type, Math. Z. 119 (1971), 149-152.

3. Introduction to Lie Algebras and Representation Theory, Graduate Texts in Math. 9, Springer 1972.

4. Projective modules for SL(2,q), J. Algebra 25 (1973), 513-518.

5. Some computations of Cartan invariants for finite groups of Lie type, Comm. Pure Appl. Math. 26 (1973), 745-755.

6. Weyl groups, deformations of linkage classes, and character degrees for Chevalley groups, Comm. Algebra 1 (1974), 475-490.

7. Representations of SL(2,p), Amer. Math. Monthly 82 (1975), 21-39.

8. Linear Algebraic Groups, Graduate Texts in Math. 21, Springer 1975.

J. E. Humphreys, D.-N. Verma

1. Projective modules for finite Chevalley groups, Bull. Amer. Math. Soc. 79(1973), 467-468.

J. C. Jantzen

1. Darstellungen halbeinfacher algebraischer Gruppen und zugeordnete kontravariante Formen, Bonner Math. Schriften 67, 1973.

2. Zur Charakterformel gewisser Darstellungen halbeinfacher Gruppen und Lie-Algebren, Math. Z. 140(1974), 127-149.

3. Darstellungen halbeinfacher Gruppen und kontravariante Formen, preprint.

4. Über das Dekompositionsverhalten gewisser modularer Darstellungen halbeinfacher Gruppen, preprint.

G. J. Janusz

1. Indecomposable modules for finite groups, Ann. of Math. 89 (1969), 209-241.

2. Simple components of Q[SL(2,q)], Comm. Algebra 1 (1974), 1-22.

A. V. Jeyakumar

    1. Principal indecomposable representations for the group SL(2,q),
       J. Algebra 30(1974), 444-458.

V. Kac, B. Weisfeiler

    1. Coadjoint action of a semi-simple algebraic group and the
       center of the enveloping algebra in characteristic p, Indag.
       Math., to appear.

M. T. Karkar, J. A. Green

    1. A theorem on the restriction of group characters, and its
       application to the character theory of SL(n,q), Math. Ann. 215
       (1975), 131-134.

N. Kawanaka

    1. On characters and unipotent elements of finite Chevalley groups,
       Proc. Japan Acad. 48 (1972), 589-594.

    2. A theorem on finite Chevalley groups, Osaka J. Math. 10 (1973),
       1-13.

    3. Unipotent elements and characters of finite Chevalley groups,
       Osaka J. Math. 12(1975), 523-554.

D. A. Kazhdan

    1. Proof of Springer's hypothesis, to appear.

R. Kilmoyer

    1. Some irreducible complex representations of a finite group with
       a (B,N) pair, Ph.D. thesis, M.I.T., 1969.

    2. The reflection character of a finite group with a (B,N) pair,
       in: Proc. Symp. Pure Math. 21, pp. 91-94, Amer. Math. Soc. 1971

M. Klemm

    1. Charakterisierung der Gruppen $PSL(2,p^f)$ und $PSU(3,p^{2f})$ durch
       ihre Charakertafel, J. Algebra 24 (1973), 127-153.

T. Kondo

    1. The characters of the Weyl group of type F_4, J. Fac. Sci. Univ.
       Tokyo, Sect. I, 11 (1965), 145-153.

116

B. Kostant

   1. Groups over $\mathbb{Z}$, in: Proc. Symp. Pure Math. 9, 90-98, Amer. Math.
      Soc. 1966.

P. J. Lambert

   1. Characterizing groups by their character tables. I, Quart. J.
      Math. Oxford Ser. 23 (1972), 427-433; II, ibid. 24 (1973), 223-
      240; III, ibid. 25 (1974), 29-40.

V. Landazuri

   1. The number of characters of Chevalley groups [Spanish], Revista
      Colombiana Mat. 6 (1972), 125-164; reviewed in Zbl. 259:20042.

V. Landazuri, G. M. Seitz

   1. On the minimal degrees of projective representations of the
      finite Chevalley groups, J. Algebra 32 (1974), 418-443.

P. Landrock

   1. A counterexample to a conjecture on the Cartan invariants of a
      group algebra, Bull. London Math. Soc. 5(1973), 223-224.

G. I. Lehrer

   1. The discrete series of the linear groups SL(n,q), Math. Z. 127
      (1972), 138-144.

   2. The characters of the finite special linear groups, J. Algebra
      26 (1973), 564-583.

   3. Discrete series and regular unipotent elements, J. London Math.
      Soc. 6 (1973), 732-736.

   4. Discrete series and the unipotent subgroup, Comp. Math. 28
      (1974), 9-19.

   5. Weil representations and cusp forms on unitary groups, Bull.
      Amer. Math. Soc. 80 (1974), 1137-1141; correction, ibid. 81
      (1975), 636.

   6. Characters, classes and duality in isogenous groups, J. Algebra
      36 (1975), 278-286.

G. I. Lehrer

7. Adjoint groups, regular unipotent elements and discrete series characters, Trans. Amer. Math. Soc. 214 (1975), 249-260.

G. Lusztig

1. On the discrete series representations of the general linear groups over a finite field, Bull. Amer. Math. Soc. 79 (1973), 550-554.

2. The discrete series of GL_n over a finite field, Ann. of Math. Studies No. 81, Princeton Univ. Press 1974.

3. On the discrete series representations of the classical groups over finite fields, Int. Congress of Math., Vancouver 1974.

4. Sur la conjecture de Macdonald, C. R. Acad. Sci. Paris 280 (1975), 317-320.

I. G. Macdonald

1. On the degrees of the irreducible representations of symmetric groups, Bull. London Math. Soc. 3 (1971), 189-192.

2. Some irreducible representations of Weyl groups, Bull. London Math. Soc. 4 (1972), 148-150.

3. On the degrees of the irreducible representations of finite Coxeter groups, J. London Math. Soc. 6 (1973), 298-300.

R. P. Martineau

1. On 2-modular representations of the Suzuki groups, Amer. J. Math. 94 (1972), 55-72.

2. On representations of the Suzuki groups over fields of odd characteristic, Bull. London Math. Soc. 4 (1972), 153-160.

S. J. Mayer

1. On the irreducible characters of the symmetric group, Advances in Math. 15 (1975), 127-132.

2. On the characters of the Weyl group of type C, J. Algebra 33 (1975), 59-67.

3. On the characters of the Weyl group of type D, Math. Proc. Cambridge Phil. Soc. 77 (1975), 259-264.

J. McKay

    1. Irreducible representations of odd degree, J. Algebra 20
       (1972), 416-418.

A. O. Morris

    1. Projective representations of Weyl groups, J. London Math. Soc.
       8 (1974), 125-133.

    2. Projective characters of exceptional Weyl groups, J. Algebra
       29 (1974), 567-586.

S. Nozawa

    1. On the characters of the finite general unitary group $U(4,q^2)$
       J. Fac. Sci. Univ. Tokyo 19 (1972), 257-293.

H. Pahlings

    1. Über die Charakterentafel der Weyl-Gruppen von Typ F_4, Mitt.
       Math. Sem. Giessen, Heft 91 (1971), 115-119.

    2. On the character tables of finite groups generated by 3-
       transpositions, Comm. Algebra 2 (1974), 117-131.

B. Pareigis

    1. Kohomologie von p-Lie-Algebren, Math. Z. 104 (1968), 281-336.

R. D. Pollack

    1. Restricted Lie algebras of bounded type, Bull. Amer. Math. Soc.
       74 (1968), 326-331.

E. W. Read

    1. Projective characters of the Weyl group of type F_4, J. London
       Math. Soc. 8 (1974), 83-93.

F. A. Richen

    1. Modular representations of split BN pairs, Trans. Amer. Math.
       Soc. 140 (1969), 435-460.

    2. Groups with a Steinberg character, Math. Z. 128 (1972),
       297-304.

    3. Blocks of defect zero of split (B,N) pairs, J. Algebra 21
       (1972), 275-279.

I. Schur

    1. Untersuchungen über die Darstellung der endlichen Gruppen durch gebrochene lineare Substitutionen, J. Reine Angew. Math. 132 (1907), 85-137.

G. M. Seitz

    1. Flag-transitive subgroups of Chevalley groups, Ann. of Math. 97 (1973), 27-56.

    2. Small rank permutation representations of finite Chevalley groups, J. Algebra 28 (1974), 508-517.

    3. Some representations of classical groups, J. London Math. Soc. 10 (1975), 115-120.

N. N. Shapovalov

    1. On a certain bilinear form on the enveloping algebra of a complex semisimple Lie algebra, Funkcional. Anal. i Prilozen. 6, no. 4 (1972), 65-70 = Functional Anal. Appl. 6(1972), 307-312.

K. Shinoda

    1. The conjugacy classes of Chevalley groups of type (F_4) over finite fields of characteristic 2, J. Fac. Sci. Univ. Tokyo 21 (1974), 133-159.

    2. The conjugacy classes of the finite Ree groups of type (F_4), J. Fac. Sci. Univ. Tokyo 22 (1975), 1-15.

T. Shoji

    1. The conjugacy classes of Chevalley groups of type (F_4) over finite fields of characteristic $p \neq 2$, J. Fac. Sci. Univ. Tokyo 21 (1974), 1-17.

A. J. Silberger

    1. An elementary construction of the representations of SL(2,GF(q)), Osaka J. Math. 6(1969), 329-338.

W. A. Simpson

    1. Irreducible odd representations of PSL(n,q), J. Algebra 28 (1974), 291-295.

W. A. Simpson, J. S. Frame

    1. The character tables for SL(3,q), SU(3,q^2), PSL(3,q), PSU(3,q^2), Canad. J. Math. 25 (1973), 486-494.

L. Solomon

    1. Invariants of finite reflection groups, Nagoya Math. J. 22 (1963), 57-64.

    2. A decomposition of the group algebra of a finite Coxeter group, J. Algebra 9 (1968), 220-239.

    3. The Steinberg character of a finite group with BN-pair in: Theory of Finite Groups, ed. R. Brauer, C. H. Sah, pp. 213-221, New York: W. A. Benjamin 1969.

    4. The affine group I. Bruhat decomposition, J. Algebra 20 (1972), 512-539.

T. A. Springer

    1. Weyl's character formula for algebraic groups, Invent. Math. 5 (1968), 85-105.

    2. Cusp forms for finite groups, in: Seminar on Algebraic Groups and Related Finite Groups, Lect. Notes in Math. 131, Springer 1970.

    3. Characters of special groups, in: Seminar on Algebraic Groups and Related Finite Groups, Lect. Notes in Math. 131, Springer 1970.

    4. Generalization of Green's polynomials, in: Proc. Symp. Pure Math. 21, pp. 149-153, Amer. Math. Soc. 1971.

    5. On the characters of certain finite groups, in: Lie Groups and Their Representations (Proc. Summer School on Group Representations of the Bolyai János Math. Soc., Budapest, 1971), pp. 621-644, New York: Halsted 1975.

    6. Caractères des groupes de Chevalley finis, Sém. Bourbaki, exp. 429, Lect. Notes in Math. 383, Springer 1974.

    7. Characters of finite Chevalley groups, in: Proc. Symp. Pure Math. 26, pp. 401-406, Amer. Math. Soc. 1973.

T. A. Springer

    8. Relèvement de Brauer et représentations paraboliques de $GL_n(\mathbb{F}_q)$, Sém. Bourbaki, exp. 441, Lect. Notes in Math. 431, Springer 1975.

    9. Trigonometrical sums, Green functions of finite groups and representations of Weyl groups, preprint.

B. Srinivasan

    1. On the modular characters of the special linear group $SL(2,p^n)$ Proc. London Math. Soc. 14(1964), 101-114.

    2. The characters of the finite symplectic group $Sp(4,q)$, Trans. Amer. Math. Soc. 131 (1968), 488-525.

    3. On the Steinberg character of a finite simple group of Lie type J. Austral. Math. Soc. 12 (1971), 1-14.

    4. Isometries in finite groups of Lie type, J. Algebra 26 (1973), 556-563.

    5. The decomposition of some Lusztig-Deligne representations of finite groups of Lie type, preprint.

    6. The characters of the finite unitary groups, preprint.

R. Steinberg

    1. The representations of $GL(3,q)$, $GL(4,q)$, $PGL(3,q)$, and $PGL(4,q)$, Canad. J. Math. 3 (1951), 225-235.

    2. Prime power representations of finite linear groups, II, Canad. J. Math. 9 (1957), 347-351.

    3. Representations of algebraic groups, Nagoya Math. J. 22 (1963), 33-56.

    4. Endomorphisms of linear algebraic groups, Mem. Amer. Math. Soc. 80 (1968).

    5. Lectures on Chevalley groups, Yale Univ. Math. Dept. 1968.

    6. On a theorem of Pittie, Topology 14 (1975), 173-177.

S. Tanaka

1. Construction and classification of irreducible representations of special linear group of the second order over a finite field, Osaka J. Math. 4 (1967), 65-84.

E. Thoma

1. Die Einschränkung der Charaktere von GL(n,q) auf GL(n-1,q), Math. Z. 119 (1971), 321-338.

F. D. Veldkamp

1. Representations of algebraic groups of type F_4 in characteristic 2, J. Algebra 16 (1970), 326-339.

D.-N. Verma

1. Structure of certain induced representations of complex semi-simple Lie algebras, Bull. Amer. Math. Soc. 74 (1968), 160-168.

2. Role of affine Weyl groups in the representation theory of algebraic Chevalley groups and their Lie algebras, in: Lie Groups and Their Representations (Proc. Summer School on Group Representations of the Bolyai János Math. Soc., Budapest, 1971), pp. 653-705, New York: Halsted 1975.

3. Modular refinement of Macdonald's conjecture for prime Chevalley groups.

W. J. Wong

1. Representations of Chevalley groups in characteristic p, Nagoya Math. J. 45 (1971), 39-78.

2. Irreducible modular representations of finite Chevalley groups, J. Algebra 20 (1972), 355-367.

T. Yokonuma

1. Sur le commutant d'une représentation d'un groupe de Chevalley fini, J. Fac. Sci. Univ. Tokyo 15 (1968), 115-129; II, ibid. 16 (1969), 65-82.

E. G. Zaslawsky

    1. Computational methods applied to ordinary and modular characters of some finite simple groups, Ph.D. thesis, U. Calif. Santa Cruz, 1974.

A. V. Zelevinskii, G. S. Narkunskaya

    1. Representations of the group $SL(2, F_q)$, where $q = 2^n$, Funkcional Anal. i Prilozen 8, no. 3 (1974), 75-76 = Functional Anal. Appl. 8 (1974), 256-257.

NOTATION

field:

K   algebraically closed field of characteristic $p > 0$

groups and Lie algebras:

G   simply connected Chevalley group over K of simple type

T   maximal torus of G

B   Borel subgroup of G containing T

U   unipotent radical of B

Γ $G(\mathbb{F}_p)$

Γ_n $G(\mathbb{F}_q)$, $q = p^n$

$K\Gamma_n$ group algebra

Γ'_n twisted group

$\underline{g}_{\mathbb{C}}$ simple Lie algebra over $\mathbb{C}$ having root system Φ

$\underline{g}$ Lie algebra of G

X_α, Y_α, H_i basis of $\underline{g}$ coming from a Chevalley basis of $\underline{g}_{\mathbb{C}}$

$\underline{t}$ Lie algebra of T

$\underline{b}$ Lie algebra of B

$\underline{n}^+$ Lie algebra of U

$\underline{u}$ restricted universal enveloping algebra of $\underline{g}$

$\underline{\underline{u}}_n$ analogue of $\underline{u}$

root system:

Φ roots of G relative to T

Φ^+, Φ^- positive (resp. negative) roots

ℓ rank

m   number of positive roots

α_i simple roots ($1 \leq i \leq \ell$)

W   Weyl group

$\ell(\sigma)$ length in W

σ_o	element of W sending Φ^+ to Φ^-
σ_i	reflection with respect to α_i
ε	alternating character of W

weights:

X	character group $X(T)$ (= full weight lattice)
X_r	subgroup of X generated by Φ
X^+	dominant weights
λ_i	fundamental dominant weights ($1 \leq i \leq \ell$)
δ	sum of all λ_i (= half sum of positive roots)
δ_σ	sum of λ_i for those i such that $\ell(\sigma_i\sigma) < \ell(\sigma)$
$\leq$	partial ordering of X
f	$[X:X_r]$
X_q	$\{\Sigma c_i\lambda_i \mid 0 \leq \rho_i < q\}$, $q = p^n$
Λ	X/pX
$\widetilde{W}$	group generated by W and translations by pX
W_a	affine Weyl group (relative to p)
$\sim$	linkage (in X or Λ)
$\lambda(\sigma)$	$\sigma(\lambda+\delta) - \delta$ in Λ
λ^o	opposite linked weight in Λ : $\sigma_o(\lambda + \delta) - \delta$

modules:

V_λ, $\bar{V}_\lambda$	irreducible $\underline{g}_{\mathbb{C}}$-module of highest weight λ ($\lambda \in X^+$) and its reduction modulo p
M_λ	irreducible G-module of highest weight ($\lambda \in X^+$)
Z_λ	universal $\underline{u}$-module of highest weight ($\lambda \in \Lambda$)
St	$M_{(p-1)\delta}$
St_n	$M_{(q-1)\delta}$, $q = p^n$
Q_λ	PIM for $\underline{u}$
R_λ	PIM for $K\Gamma$
z_i, $\bar{z}_i$	irreducible $\mathbb{C}\Gamma$-module and its reduction modulo p

formal characters:

$\mathbb{Z}[X]$	group ring of X
$e(\lambda)$	formal symbols $(\lambda \in X)$, basis of $\mathbb{Z}[X]$
$m_V(\mu)$	multiplicity of μ in V
$ch(V)$	$\Sigma m_V(\mu)e(\mu)$
$ch(\lambda)$	$ch(V_\lambda)$
$p\text{-}ch(\lambda)$	$ch(M_\lambda)$
$dim(\lambda)$	$dim(V_\lambda)$
$p\text{-}dim(\lambda)$	$dim(M_\lambda)$
$s(\mu)$	sum over W-orbit of $e(\mu)$
$s'(\mu)$	sum over W-orbit of μ mod $(p-1)X$, viewed as a Brauer character

constants:

a_λ	cardinality of linkage class of λ
$a_{\lambda\mu}$	$p\text{-}ch(\lambda) = \Sigma a_{\lambda\mu}ch(\mu)$
$b_{\lambda\mu}$	$ch(\lambda) = \Sigma b_{\lambda\mu}p\text{-}ch(\mu)$
$c_{\lambda\mu}$	Cartan invariants
d_λ	decomposition numbers: $Z_\mu \underset{u}{\leftrightarrow} \Sigma d_\lambda M_\lambda$

INDEX

affine Weyl group, 13

alcove, 14, 65

Ballard's thesis, 55

block, 59, 100

Brauer complex, 65, 82, 89

Brauer tree, 59

Cartan invariants, 29, 57, 101

decomposition matrix, 101

decomposition number, 23

deformation of linkage class, 67, 80

dominant weight, 5

Ennola conjecture, 99

formal character, 7

Frobenius algebra, 100

fundamental dominant weight, 5

fundamental group, 5

G-module, 5

intertwining operators, 27

linkage principle, 11

linked PIM's, 100

linked weights, 11

maximal vector, 6

minimal vector, 6

Mumford conjecture, 43

negative root, 5

p-regular weight, 14

PIM, 21, 100

positive root, 5

principal indecomposable module, 21, 100

ramification, 26

regular weight, 52

regularity conjecture, 52

restricted weight, 6

root system, 5

special point, 82

Steinberg module, 9, 50, 102

symmetric algebra, 101

twisted group, 95

twisted tensor product, 8, 38

Verma conjectures, 17, 31

Verma module, 21, 22

weight, 5

Weyl group, 5

Weyl module, 7

Vol. 399: Functional Analysis and its Applications. Proceedings 1973. Edited by H. G. Garnir, K. R. Unni and J. H. Williamson. II, 584 pages. 1974.

Vol. 400: A Crash Course on Kleinian Groups. Proceedings 1974. Edited by L. Bers and I. Kra. VII, 130 pages. 1974.

Vol. 401: M. F. Atiyah, Elliptic Operators and Compact Groups. V, 93 pages. 1974.

Vol. 402: M. Waldschmidt, Nombres Transcendants. VIII, 277 pages. 1974.

Vol. 403: Combinatorial Mathematics. Proceedings 1972. Edited by D. A. Holton. VIII, 148 pages. 1974.

Vol. 404: Théorie du Potentiel et Analyse Harmonique. Edité par J. Faraut. V, 245 pages. 1974.

Vol. 405: K. J. Devlin and H. Johnsbråten, The Souslin Problem. VIII, 132 pages. 1974.

Vol. 406: Graphs and Combinatorics. Proceedings 1973. Edited by R. A. Bari and F. Harary. VIII, 355 pages. 1974.

Vol. 407: P. Berthelot, Cohomologie Cristalline des Schémas de Caracteristique p > o. II, 604 pages. 1974.

Vol. 408: J. Wermer, Potential Theory. VIII, 146 pages. 1974.

Vol. 409: Fonctions de Plusieurs Variables Complexes, Séminaire François Norguet 1970–1973. XIII, 612 pages. 1974.

Vol. 410: Séminaire Pierre Lelong (Analyse) Année 1972–1973. VI, 181 pages. 1974.

Vol. 411: Hypergraph Seminar. Ohio State University, 1972. Edited by C. Berge and D. Ray-Chaudhuri. IX, 287 pages. 1974.

Vol. 412: Classification of Algebraic Varieties and Compact Complex Manifolds. Proceedings 1974. Edited by H. Popp. V, 333 pages. 1974.

Vol. 413: M. Bruneau, Variation Totale d'une Fonction. XIV, 332 pages. 1974.

Vol. 414: T. Kambayashi, M. Miyanishi and M. Takeuchi, Unipotent Algebraic Groups. VI, 165 pages. 1974.

Vol. 415: Ordinary and Partial Differential Equations. Proceedings 1974. XVII, 447 pages. 1974.

Vol. 416: M. E. Taylor, Pseudo Differential Operators. IV, 155 pages. 1974.

Vol. 417: H. H. Keller, Differential Calculus in Locally Convex Spaces. XVI, 131 pages. 1974.

Vol. 418: Localization in Group Theory and Homotopy Theory and Related Topics. Battelle Seattle 1974 Seminar. Edited by P. J. Hilton. VI, 172 pages 1974.

Vol. 419: Topics in Analysis. Proceedings 1970. Edited by O. E. Lehto, I. S. Louhivaara, and R. H. Nevanlinna. XIII, 392 pages. 1974.

Vol. 420: Category Seminar. Proceedings 1972/73. Edited by G. M. Kelly. VI, 375 pages. 1974.

Vol. 421: V. Poénaru, Groupes Discrets. VI, 216 pages. 1974.

Vol. 422: J.-M. Lemaire, Algèbres Connexes et Homologie des Espaces de Lacets. XIV, 133 pages. 1974.

Vol. 423: S. S. Abhyankar and A. M. Sathaye, Geometric Theory of Algebraic Space Curves. XIV, 302 pages. 1974.

Vol. 424: L. Weiss and J. Wolfowitz, Maximum Probability Estimators and Related Topics. V, 106 pages. 1974.

Vol. 425: P. R. Chernoff and J. E. Marsden, Properties of Infinite Dimensional Hamiltonian Systems. IV, 160 pages. 1974.

Vol. 426: M. L. Silverstein, Symmetric Markov Processes. X, 287 pages. 1974.

Vol. 427: H. Omori, Infinite Dimensional Lie Transformation Groups. XII, 149 pages. 1974.

Vol. 428: Algebraic and Geometrical Methods in Topology, Proceedings 1973. Edited by L. F. McAuley. XI, 280 pages. 1974.

Vol. 429: L. Cohn, Analytic Theory of the Harish-Chandra C-Function. III, 154 pages. 1974.

Vol. 430: Constructive and Computational Methods for Differential and Integral Equations. Proceedings 1974. Edited by D. L. Colton and R. P. Gilbert. VII, 476 pages. 1974.

Vol. 431: Séminaire Bourbaki – vol. 1973/74. Exposés 436–452. IV, 347 pages. 1975.

Vol. 432: R. P. Pflug, Holomorphiegebiete, pseudokonvexe Gebiete und das Levi-Problem. VI, 210 Seiten. 1975.

Vol. 433: W. G. Faris, Self-Adjoint Operators. VII, 115 pages. 1975.

Vol. 434: P. Brenner, V. Thomée, and L. B. Wahlbin, Besov Spaces and Applications to Difference Methods for Initial Value Problems. II, 154 pages. 1975.

Vol. 435: C. F. Dunkl and D. E. Ramirez, Representations of Commutative Semitopological Semigroups. VI, 181 pages. 1975.

Vol. 436: L. Auslander and R. Tolimieri, Abelian Harmonic Analysis, Theta Functions and Function Algebras on a Nilmanifold. V, 99 pages. 1975.

Vol. 437: D. W. Masser, Elliptic Functions and Transcendence. XIV, 143 pages. 1975.

Vol. 438: Geometric Topology. Proceedings 1974. Edited by L. C. Glaser and T. B. Rushing. X, 459 pages. 1975.

Vol. 439: K. Ueno, Classification Theory of Algebraic Varieties and Compact Complex Spaces. XIX, 278 pages. 1975

Vol. 440: R. K. Getoor, Markov Processes: Ray Processes and Right Processes. V, 118 pages. 1975.

Vol. 441: N. Jacobson, PI-Algebras. An Introduction. V, 115 pages. 1975.

Vol. 442: C. H. Wilcox, Scattering Theory for the d'Alembert Equation in Exterior Domains. III, 184 pages. 1975.

Vol. 443: M. Lazard, Commutative Formal Groups. II, 236 pages. 1975.

Vol. 444: F. van Oystaeyen, Prime Spectra in Non-Commutative Algebra. V, 128 pages. 1975.

Vol. 445: Model Theory and Topoi. Edited by F. W. Lawvere, C. Maurer, and G. C. Wraith. III, 354 pages. 1975.

Vol. 446: Partial Differential Equations and Related Topics. Proceedings 1974. Edited by J. A. Goldstein. IV, 389 pages. 1975.

Vol. 447: S. Toledo, Tableau Systems for First Order Number Theory and Certain Higher Order Theories. III, 339 pages. 1975.

Vol. 448: Spectral Theory and Differential Equations. Proceedings 1974. Edited by W. N. Everitt. XII, 321 pages. 1975.

Vol. 449: Hyperfunctions and Theoretical Physics. Proceedings 1973. Edited by F. Pham. IV, 218 pages. 1975.

Vol. 450: Algebra and Logic. Proceedings 1974. Edited by J. N. Crossley. VIII, 307 pages. 1975.

Vol. 451: Probabilistic Methods in Differential Equations. Proceedings 1974. Edited by M. A. Pinsky. VII, 162 pages. 1975.

Vol. 452: Combinatorial Mathematics III. Proceedings 1974. Edited by Anne Penfold Street and W. D. Wallis. IX, 233 pages. 1975.

Vol. 453: Logic Colloquium. Symposium on Logic Held at Boston, 1972–73. Edited by R. Parikh. IV, 251 pages. 1975.

Vol. 454: J. Hirschfeld and W. H. Wheeler, Forcing, Arithmetic, Division Rings. VII, 266 pages. 1975.

Vol. 455: H. Kraft, Kommutative algebraische Gruppen und Ringe. III, 163 Seiten. 1975.

Vol. 456: R. M. Fossum, P. A. Griffith, and I. Reiten, Trivial Extensions of Abelian Categories. Homological Algebra of Trivial Extensions of Abelian Categories with Applications to Ring Theory. XI, 122 pages. 1975.